W9-DEW-899

HIGH BOATS

HIGH BOATS

A Century of Salmon Remembered

Pat Wastell Norris

HARBOUR PUBLISHING

Dedicated to Barrie McClung and David Huson

Published by
Harbour Publishing Co. Ltd., P.O. Box 219, Madeira Park, BC V0N 2H0
www.harbourpublishing.com

Cover and page design by Li Eng-Lodge
Design Assistant Felicia Lo
Cover photo by Brian Gauvin

Printed and bound in China through Colorcraft Ltd., Hong Kong

Harbour Publishing acknowledges financial support from the Government of Canada through the Book Publishing Industry Development Program and the Canada Council for the Arts; and from the Province of British Columbia through the British Columbia Arts Council and the Book Publisher's Tax Credit through the Ministry of Provincial Revenue.

National Library of Canada Cataloguing in Publication Data

Norris, Pat Wastell
High boats : a century of salmon remembered / Pat Wastell Norris.

Includes bibliographical references and index.
ISBN 1-55017-289-1

1. Salmon fisheries—British Columbia—Pacific Coast—History. 2. Pacific Coast (B.C.)—History. I. Title.
SH349.N67 2003 338.3'72756'097111 C2003-911084-2

TABLE OF CONTENTS

ACKNOWLEDGEMENTS

Credit for this book must go to Barrie McClung whose scraps of conversation inspired it. Barrie has a passion for history and an unbeatable memory for detail. He not only provided pages of information, including a comprehensive description of the seining process, but also introduced me to David Huson. On the strength of that introduction David gave me access to his family's documents and photographs (a remarkable record) and shared his own story. Thank you both for making this book possible. Thank you, too, to those invisible women, Linda and Susan, who accepted their exclusion with such good grace.

Then come the others. Some of the things you told me were hilarious—some were heartbreaking. In alphabetical order you are:

Mike Acres; Alert Bay Public Library's Joyce Wilby; Michael Berry; the BC Salmon Farmers' Association's Gina Forsyth-Dodd and Anita Peterson; the CBC's Kim Belle; George and Ruth Cook; Department of Justice librarian Isabel Schulthess; Fisheries and Oceans Canada librarians Louise Archibald, Joyce Kennedy, Susan Spohn and Marcia Vanwely; Rod Hourston; Lorraine Hunt; Ken Lands; Museum at Campbell River curator Linda Hogarth; Tiiu McCormick; Don Millerd; Alexandra Morton; Don Pepper; David Rahn, former editor of *Fisherman Life*; Grace and Julian Stauffer; Kirk Stinchcombe; the David Suzuki Foundation; Ian Todd; the UFAWU-CAW's Bruce Logan; the U'Mista Cultural Centre's Juanita Pasco; John Volpe; Vancouver Maritime Museum librarian Rachael Grant; Steve Warren; Judy Warren; Mike Weigold.

Thank you all for sharing your knowledge and your experience so generously.

The writing sequence puts the editing at the end but this in no way diminishes my debt to two great editors, Shane McCune and Andrew Scott.

INTRODUCTION

"I think I would be lost if I ever quit what I was doing because I started very young. I would like to keep on fishing because that is my life. I like fishing and I don't know if I am qualified to fit into anything other than fishing."
— Jimmy Sewid, *Guests Never Leave Hungry: The Autobiography of James Sewid, a Kwakiutl Indian*

Steve Warren was born in Alert Bay in 1914 when the village was still dominated by Kwakwaka'wakw longhouses. He died just before this book was finished. By that time he had lost his hearing and his sight, he was in a wheelchair and a dialysis machine had taken over his kidneys' function. At 88 everything had failed him but his mind.

Until he was 13, Steve spent fishing season at the lagoon on Harbledown Island, where his mother was a fish buyer. I grew up across Johnstone Strait in a tiny community called Telegraph Cove. We had a lot in common, and in his cluttered living room in Oak Bay we talked about what it was like to grow up on a coast that was still remote. When he told me about Harbledown I was intrigued by the idea of a lagoon. Living where the tide was trapped for a few hours each day suggested all kinds of possibilities not available to a child growing up where the tides swept past and the shore plunged down to unknowable depths.

"Did you ever swim in the lagoon?" I yelled into Steve's hearing aid.

Steve had a soft deep voice, and he always spoke slowly and deliberately. "No," he said. "I'm not much of a swimmer."

Nor was I. None of the north coast's inhabitants were swimmers. Although we lived within spitting distance of the sea, and spent most of our lives afloat on it, we weren't swimmers. In half an hour that frigid water killed people. We were either aboard—or in terminal trouble. About once a month someone, somewhere in the area, drowned. A few of them were toddlers, a few were drunks and the rest were just earning their living. Once in a very great while they were lucky. The crew of a boat that sank at the entrance to Knight Inlet made it to shore "with the assistance of driftwood."

The place where I grew up—the place where the water was cold enough to kill you—was the "North Island," the northerly half of Vancouver Island. The road stopped at Campbell River and the climate changed at Seymour Narrows; it was a wild, wet wilderness, a world away from the towns and cities and arbutus trees on the southern half of the island. My family lived in a small isolated cove on the northeast end of Vancouver Island where seeing a doctor, shopping, banking or visiting my grandparents—in fact, connecting with the outside world in any way at all—involved a trip to Alert Bay by boat. To do any of these things we started our tug's heavy-duty diesel engine, let go the lines and headed out into Johnstone Strait. On a bright summer day with the sea glittering in the sun, the trip to Cormorant Island was stunningly beautiful. In winter's southeasters, though, with waves heaving and hissing around us and water smashing against the wheelhouse windows and pouring down the deck, it was not a trip for the faint of heart.

Sometimes, before I was old enough to go to school, I spent days with my grandparents in Alert Bay. In the afternoons, after my grandmother had scrubbed clothes, cooked a lot of delicious food on a wood stove, washed and trimmed the oil lamps and done her bookkeeping (for she was the family financial officer), she changed into a fresh pair of lisle stockings, put on a navy blue "afternoon dress," did up her hair and we went out. I took her hand and together we walked along the wide gravel path that followed the shoreline. What made these excursions rather different from the average was

OPPOSITE: Boys hang out on the Alert Bay waterfront, throwing stones and generally looking for mischief, circa 1920. From the collection of the author

the fact that our route led us right through a Kwakwaka'wakw First Nation village. On one side of the path there was a row of longhouses fronted by towering totem poles, and we walked beneath the cruel proud gaze of thunderbirds and passed bears that exhibited their ferocious teeth. We developed a nodding acquaintance with huge naked figures with pursed lips, extended arms and sightless eyes that stared out to sea. On the opposite side of the path there were dugout canoes pulled up on the pebbly beach and, in season, big beach fires ringed with intricate racks of drying salmon. On these excursions we met barefoot children clutching tiny kittens with crusted eyes and we dodged half-starved dogs scratching and nipping at their limitless supply of fleas. Neither my grandmother nor I saw anything noteworthy about these surroundings. She had lived in Alert Bay since 1909 and I had never known anything else.

It was probably just as well that I hadn't, for Alert Bay offered other, more unsettling experiences for a small girl from a virtually teetotal household. If our boat was tied to one of the docks in "downtown" Alert Bay on a Friday or Saturday afternoon, and if I accompanied my father up the street on an errand, I had to edge carefully around the drunks who lay sprawled on the side of the road mumbling and shouting (fortunately unintelligible) curses. Inevitably their faces were covered with blood, the result of a difference of opinion with their fellows.

"The Bay is a terrible place," said my mother. She was dead against drink.

A lot of the people like us, who lived in outlying communities, earned their living in the forest industry, but the people of Alert Bay were fishermen. The docks along the crescent of the bay were jammed with their boats: seiners, trollers and gillnetters. A man and his boat were thought of as an inseparable unit. Nothing that happened to one would not involve the other. And a man with his own boat had stature in the community because it meant he was hardworking and responsible and a good seaman.

There were good fishing seasons and poor fishing seasons but by the time the dirty thirties arrived it hardly mattered. After 1929 there were still plenty of fish around but nobody seemed to be buying them. In the cities men were lining up at soup kitchens, where they didn't serve salmon. For a year or two this could have been considered tough luck but it lasted much longer than a couple of years; it ground on for more than 10. In 1933 the prices for fish were so low that Steve Warren went over to Englewood to get a job in the sawmill there. He worked for two weeks before the mill closed for Christmas and for those two weeks he earned a total of eight dollars (plus room and board, of course).

We were a populace that worked brutally hard and never had so much as a whiff of leisure or luxury. It's important to understand that when you think about what came afterward and why we embraced it so heedlessly. It's important to see the present in relation to the past. It's foolish, for example, to get on a high horse with the salmon farmers when they shoot seals. They aren't the first. In 1951 Alert Bay's *Pioneer Journal* reported that the fisheries patrol boats had shot more than 100 animals in the annual seal hunt.

It was World War II that finally caused fish prices to rise and brought a kind of prosperity. In short order there were lots of jobs available in the forest industry, all of which paid seemingly enormous wages, and benefits to boot. We got hydro and a road that connected us with the rest of the world; small businesses had more customers, and they paid cash; and the fishing industry got a lot of new technology and made more money for its fishermen than it had ever done before.

After so many years of privation we couldn't believe our luck. Not being used to having discretionary income, some people saved it and some blew it as fast as it came in. My parents bought themselves a record player, and my mother sent for a selection of classical recordings offered by *Reader's Digest*. Life, after all our struggles, was so good that none of us noticed the changes taking place.

But they *were* taking place. Things were different back when Steve was born. "The early sockeye fishing at the Nimpkish was very interesting," he said, not long ago. "It's a lot different today." Then he caught himself. "Well, there isn't any today," he added. The white people's Bible warned that the seven good years are followed by the seven bad. The First Nations' First Salmon rite taught that nature's abundance must never be taken for granted. But somehow this ancient wisdom got lost and everyone let the good times roll.

I have some women friends who, although they wish me well, never read any of my books. They were gently reared and I was not, and they object to the "bad language" they find therein. It wasn't my parents' fault that I wasn't reared gently. They did the best they could, but circumstances were against it. All the comfortable barriers that shield the middle class from the nastiness of real life were missing in a place like the North Island, so life was right there in your face. Of course, humanity's best side was there, too. At a hospital meeting in Alert Bay, the doctor stood up and thanked the patrons of the two local beer parlours for being "a never failing source of willing blood donors." And as a child in the 1950s, David Huson had no idea that there was something called racial discrimination.

For the convenience of the gently reared I have put the first swear word in the second sentence of this book so they can tell right away that it isn't a book for them. But those who can get past the word "shit" will find that this story chronicles the fortunes of the Nimpkish River salmon runs over 100 years and more. These are the fish that David Rahn, former editor of *Fisherman Life* magazine, considers unparalleled. "The Johnstone Strait sockeye are still the best-tasting salmon in the whole world," he says.

We don't seem to have appreciated that fact. First we squandered them, catching more than canneries could handle and dumping the rest in the sea. Then we sprayed their spawning grounds with DDT. Then we clear-cut the forests around those spawning grounds. Now we are unleashing the unknown dangers of fish farming upon them. In the process we are destroying a way of life as well as several species of fish.

When David Huson was growing up, Alert Bay was a rip-roaring little village full of seine boats and shopkeepers and drunks and Anglicans and fishermen. It had a community hall built with volunteer labour and a doctor who flew his own plane and played jazz piano. Now it has lost its commerce and its reason for being, and people cling to it as they would to a drifting raft.

In the 1950s and '60s, in fishing season, the *Pioneer Journal*, the local newspaper, always contained a box on its front page entitled "High Boats." These were the boats with the highest catch for the week. The boats' names, their skippers and the fishing companies they worked for were listed. Some boats—the ones with the skippers who were highliners—made the list week after week; some fluked it; some never got their names in the paper. Yet week after week they were all out there—down Johnstone Strait, in Blackfish Sound, off Malcolm Island—setting their nets and hoping to make it big. There are no high boats now. Before long there may be no more wild salmon.

The trouble is, man is probably the most invasive species on the planet, spreading over the world like crabgrass. Even the least rapacious among us want a comfortable, healthy, enjoyable life and enough money to ensure this for our families. The difficulty lies in our ever-increasing numbers. We demand more and more fresh water, timber, electricity—and food. So unlikely places like Alert Bay, a remote little village surrounded by a scattering of exquisite islands and fronted by a long deep strait, are impacted by wave after wave of change.

People like Bill Cranmer, whose genetic links go back for thousands of years, and David Huson, whose great-grandfather came here over a century ago, and Billy Proctor, who has spent all his life in the area, and Alexandra Morton, who after 20 years is a relative newcomer, all love this place with a passion that is not always easily expressed. They function best here. How long they will continue to do so is the question.

CHAPTER 1

High Boats:
Barclay Sound — Fred Joliffe (BCP)
Twin Sisters — James Sewid (ABC)

PIONEER JOURNAL, OCT. 29, 1952

David summed it up. "It's all gone to rat shit," he said.

"Well, at least *you're* still hanging in there," said Barrie. "You're still alive and kicking. Better a live dog than a dead lion."

"Where do you get this stuff?"

"The Bible," said Barrie. "That's what it says in the Bible. My grandfather read it to us every Sunday. Before dinner."

David was not a theologian. He flipped his cigarette butt over the side into the water. "Anyway," he said, "it's a real good day."

"Ah . . . ik. Ik'nala."

Dave grinned. "You haven't forgot your Indian?"

"I haven't forgotten anything," said Barrie.

David Huson and Barrie McClung hadn't seen each other for years. And then, one September day, the phone rang and it was David. "You want to come up to the Bay and help me bring my boat down?" he asked.

Barrie took a moment to think. "Okay," he said. "Okay, sure."

With a ferry connection it's possible to drive to Alert Bay now, a circumstance especially appreciated by old-timers, who remember a 22-hour voyage if the weather was good and a longer one if it wasn't. Barrie got a ride up-island with David's sister, Grace Stauffer, and her husband, Sonny. Grace drove because, like many of the First Nations people, her husband suffers from diabetes. The disease has cost him a leg and he finds its replacement a burden. They had barely begun their journey before he reached up his pant leg, unbuckled his artificial limb and threw it on the dashboard. That prosthesis, complete with sock and shoe, lying there across the dashboard of the van epitomized the transition all three of them were making—from Victoria, where appearance counted, to Alert Bay, which is pretty rangitang and always has been.

Now Barrie and David sat on the *May S.*'s bulwarks outside her galley door, and behind them Alert Bay basked in unaccustomed sunshine.

The village of Alert Bay follows the long curve that makes up the western side of Cormorant Island. Up until the 1960s it was the commercial centre for a large amorphous area that spread out between Vancouver Island and the mainland. "The Hub of the Island," its little newspaper boasted. Alert Bay's customers arrived by water across the surrounding channels, straits and sounds, and its supplies arrived by water as well. Cormorant Island, shaped rather like a recumbent jelly bean, lies at the northern end of Johnstone Strait, one of the main thoroughfares for shipping in the Inside Passage, so it was ideally situated and bustlingly busy.

Of course, in most people's minds, the absence of roads and their easy communication renders a place remote, if not invisible. The population of the Lower Mainland had the vague idea that Vancouver Island stopped somewhere around Campbell River, which coincidentally was where the highway stopped. What lay beyond was anybody's guess. So Alert Bay, busy as it was, remained the preserve of loggers, fishermen and mariners. It welcomed its water-borne customers with a string of long docks: four oil docks, a couple of cannery docks and the hospital dock. And then the canneries closed and the highway battled its way to the northern tip of Vancouver Island and that changed everything for Alert Bay. Communities clustered where there was road access, supplies came and went by truck, and a village on an island was out of the loop. Alert Bay's

Although it was never a conventional 'company town,' Alert Bay shares the utilitarian frame of mind—and construction—common to resource towns up and down the West Coast.

docks, formerly swarming with boats, were demolished so that only stubs, like amputated limbs, remained. Even the hospital dock, a destination desperately sought out by countless boats carrying smashed loggers on stretchers, women in the throes of childbirth and children feverish with pneumonia, was ripped down. It was so much easier now to drive to the hospital in Port Hardy, especially on winter nights when the southeasters raged in Johnstone Strait.

In an area of great natural beauty, Alert Bay isn't—and never was—a beautiful town, nor does this make it unique. By and large the towns and villages of the north coast are an affront to the landscape that surrounds them. Many were built to service the logging and fishing industries, thrown up quickly by big resource-based companies, and they are unremittingly utilitarian. Coastal community boundaries are defined by the sea in front and a solid mass of forest behind, and they straggle along the shoreline in a scattering of floats, docks, warehouses, flimsy frame buildings and Atco trailers. Where the townsites needed level land, bulldozers have torn away the edges of the forest.

At least the village of Alert Bay has the distinction of having grown up over a period of more than 100 years, which is time immemorial by local standards. Its age, and the fact that it isn't a "company town," have given it certain architectural distinctions: the square brick block that was an Anglican residential school, a delicate white gingerbread church and the remains of a Kwakwaka'wakw village. Still, like all its neighbouring communities, it reflects a life made up of back-breaking work and hellraising. Such a life pretty well eliminates aesthetic considerations, which are dimly understood and little valued. This was clearly demonstrated in 1958, when the Alert Bay village council decided to celebrate BC's centennial by engaging in a "cleanup/paint-up" program. Appealing to the residents' pride, they exhorted them to spruce the place up a bit by painting their houses. And a few citizens were stirred to action. They bought cans of paint and painted their houses—but only the fronts. This may have had something to do with the fact that, traditionally, only the front of a longhouse was decorated, but it wasn't what the village council had in mind.

Despite the fact that the docks have shrunk to two or three, there is still ample moorage for the diminished fishing fleet, for Alert Bay has two breakwaters, each built to serve a specific segment of the population and each funded by a different federal government agency. Behind the "Indian" breakwater, so named because it was funded by the Department of Indian Affairs, the *May S.* lay moored at a float.

In the 1970s, boat builders began to build aluminum seiners. With their high bows, raked-forward wheelhouses and bristling antennae they looked like the aggressive fish hunters they were. The *May S.*, however, was a wooden boat from an earlier era and her lines were gentler. To the cognoscenti her provenance was unmistakable. She was a classic seine boat built in the 1920s. Her lines revealed a skilled boat builder and her condition, a conscientious owner. The paint work on her hull and superstructure was an immaculate white, and her varnished trim showed none of the ravages that salt air exacts. Right down her length ran a wide stripe of turquoise, and the same colour was picked up on the narrow strip of trim on her cabin roof and shadowed the big clear letters of her name. "*May S.*," read the block letters on her bow and across the front of the dodger. The dodger, that seine boat equivalent of a flying bridge, had a rakish modern canopy behind it that sprouted the antennae for her electronics. Her lines were neatly coiled and secured; her gear—even her deck mop—was fastened in place. On the afterdeck the seine drum, a 2.5-metre aluminum spool, was fitted onto a metal extension that stretched out over the stern. The skiff was pulled up onto this and secured to the boom. Everything about her proclaimed her status as a workboat, and everything about her disclosed a proud and seamanlike owner. Only the white plastic chairs up on the dodger revealed that she was also a boat for pleasure—a boat to be enjoyed. Now past her wooden hull the strong tide slipped by, carrying bits of seaweed and the occasional beer can, and the water, swept clean by this tide, revealed every pebble—every barnacle and sea anemone—in its glittering depths.

Except for an outboard motor, cranked up to the max, whining in the distance, there was no sight or sound of human activity on this early morning. Only the two men, who finally left the sunshine on deck, went into the galley and began to unpack the cartons of groceries on the galley table. Even on this bright day the interior of the boat was damp and chill. The smell of diesel oil and years of fish slime had infiltrated the boat's very bones, and the odour oozed into the cold air and penetrated clothing and bedding and food.

The *May S.* was built in Steveston in the days before that port became a suburb of Vancouver—before it was sanitized and gentrified and the streets were filled with generic day trippers surging from one T-shirt shop to the next. It was a real town then—a working town—and its work was fish. The one main street followed the contours of the riverbank; on one side weather-beaten wooden buildings housed ship chandleries, propeller adjusters, grocery stores and a Chinese restaurant. On the river side there were canneries, an ice plant and boatyards. The *May S.* was built in one of these boatyards—carefully, stoutly built of fine first-growth boat lumber.

She was powered by a big, heavy-duty, 85-horsepower Atlas

Imperial. These engines were virtually indestructible and their weight gave seiners stability. But when high-speed engines with more power and less bulk appeared in the late 1950s, the *May S.* was re-engined. She was now powered by a Volvo Penta; its compact size made room for a second diesel that drove a seine drum, spooling gear, winches and the bow thruster. To compensate for the weight of the Atlas, 37 tracks from a derelict shipyard had been placed under the Penta, along with a tonne of lead purchased from the BC Ferry Corporation. And in the hatch, under the destabilizing height of the seine drum, was a tonne of concrete.

Groceries stowed, the men slid down the companionway into the engine room and surveyed this machinery. David checked the batteries and pumped up the "day tank" for the galley stove; Barrie checked the coolant, engine oil and clutch oil levels. Satisfied, they went on deck and David swarmed up the ladder onto the roof of the wheelhouse with an agility that belied his solid bulk. Like a cowboy on his horse, he and the *May S.* were a single entity. Countless times he had stood up there and put the wheel over, sending the *May S.* into an arcing turn, her seine net spewing out behind her. Countless times he had made a set, wondering each time, like the players in Vegas, if he would hit the jackpot. From up there on the dodger he had a clear view ahead and astern. He could see the skiff man and the deck crew waiting tensely below, wondering, like him, what riches lay below the surface of the water—determined, like him, not to screw up and blow an opportunity. There was no net sliding over the stern on this day, however. He pushed the starter button and the Penta sprang to life. Just for a moment he stood motionless on the dodger, a thickset man with a broad pleasant face, eyes permanently crinkled at the corners from staring at sunlit water. Behind him the exhaust puffed away.

Nearly three months before, on the first of July, David had been on his way from Victoria to buy fish in Bella Coola. He got as far as Alert Bay. When he checked the latest prices he found that the Americans were buying chum salmon in Alaska for 33 cents a kilogram. "I figured I wouldn't even make enough to pay my crew," he said, so he just left the boat where it was and went back home to Victoria.

May. S.

David Huson at the wheel of the May S. Mike Weigold photo

PREVIOUS PAGE: *To the discerning eye, the graceful lines of the* May S. *instantly identify her as a classic seiner of the 1920s.* Vance Hanna photo

The *May S.* had been there behind the breakwater ever since, and by this day in September other vessels, packed tightly alongside and astern, had boxed her in. David spun the engine-room controls and a burst of exhaust flew away from the stack. He turned the wheel just enough and the big boat moved forward. He spun the controls again, moving from clutch to throttle with the deft grace of the drummer in a rock band. He backed, angled, backed again while Barrie clambered over the other boats slacking off, untying, hauling in and retying a maze of head and stern lines. Without a word between them they extricated the boat and she curved astern and out into the bay. David turned the wheel, eased in the clutch and gave her a bit of power. She moved ahead and they slipped past truncated docks and the hollow remains of a former cannery—the remnants of a dying, if not already dead, community. Then David pointed the boat's nose out into the strait. Lining up the forward stay with a position on a far mountain-side, he dug in his shirt pocket for a battered package of cigarettes. Then, correcting his course with his foot, he pulled out a cigarette, lit it, took a deep drag and slowly exhaled. The voyage of the *May S.* had begun.

Johnstone Strait is a long sombre stretch of water roughly five kilometres across and 400 metres deep. It follows the contours of Vancouver Island in an east-west direction for some 65 kilometres, and all along its western edge the thickly forested mountains of the Island drop straight down into the sea. Periodically this wall of mountain slopes is cleft by a deep river valley that stretches into even wilder country and ends in a jumble of snowcapped peaks. To the east of Johnstone Strait, the profiles of one island against another stretch away for kilometres to the line of mountains that marks the mainland. On gloomy days when the great humps of mountains on Vancouver Island are hung with cloud and the sea is like chipped grey granite, there are only hints of the strait's menace. But in a storm-force southeaster that menace is unmistakable. When the wind meets the strait's strong tides the seas become short, steep and ugly. Then an unending procession of foam-streaked combers rolls up the strait and their breathy roar is fair warning to mariners.

On this perfect day in September, however, the long straight stretch of water was the improbable blue of a bottle of mouthwash. Over on the *May S.*'s port side the distinctive white cone of Mount Waddington, the highest peak in BC's Coast Mountains, rose up into the sky like a western Fujiyama.

Barrie had lit the oil stove and made coffee. Now he climbed the ladder with two mugs clutched in his free hand and they stood there on the dodger with the cold briny air streaming by, sipping coffee and watching the *May S.* cut through the water. David's big gut may have said one thing but the massive forearm on the wheel said quite another. "David's always been one of those really strong guys," said a former wheel partner. "We call him Big D." Barrie, on the other hand, was slimmer than his friend and taller. And where David's brown hair was a tumble of neat short curls, Barrie's was fair and thinning. Their work shirts were worn and their jeans cut for comfort. Up top there, with the boat heeling over in the tide, they were at home.

"We'll go down to Robson Bight," said David, "and have a look at the whales."

There have always been killer whales here. East of Malcolm

Island, Blackfish Sound is named for them. They are huge creatures, the males averaging almost eight metres long and weighing eight tonnes. They roam the area in pods, travelling through the passages and rolling up and down Johnstone Strait in search of fish. But Robson Bight is unique in that the northern resident whales have made it their "rubbing beach." Groups of them gather, squeaking and squealing and roiling the water as they scrape their immense bodies against the pebbly bottom. Most intriguing is the fact that there are no rubbing beaches south of Seymour Narrows and none, that marine mammalogist John Ford has heard of, in Alaska. Transient whales do not stop there. Robson Bight, it appears, is unique; the gathering place for locals—the whale equivalent of Cheers. For they come not only to rub themselves but to socialize, and on these occasions their squeaks and squeals and whistles are so distinctive that, without being told, Ford can identify them as socializing sounds made at the rubbing beach.

To the early Kwakwaka'wakw people, whales were a symbol of power and mystery. They were the chiefs of the dolphins and porpoises and creatures that could breach two realms—air and water. Like the Kwakwaka'wakw, the blackfish live in a very structured and close-knit community, with the pods that form a "clan" each using its own distinctive communication dialect. And like the Kwakwaka'wakw, the whales are salmon eaters. But there was salmon enough for all in those early days.

The First Nations people were, as the name implies, the coast's first residents. They had lived on its bounty for thousands of years. Then Yugoslavs, Finns, Englishmen, Scots, Norwegians, Chinese, Japanese—people from the ends of the earth—found their way to the coast. They became loggers and fishermen, proprietors of restaurants and grocery stores, and owners of boat ways and sawmills and beer parlours. Without exception these people valued independence and were willing to pay the price that independence exacts, which meant that they were, above all else, practical. In the 1950s and '60s, when educated people began to arrive for the express purpose of studying the blackfish, this seemed to the locals an eccentric and frivolous way of life.

The locals were as aware of the whales in the sea as they were of the cougars in the woods. But these new people brought post-war technology with them—scuba-diving gear and underwater cameras and audio equipment. Now the squealing conversations of the whales were being recorded, their pods counted, their habits studied. To the astonishment of long-time residents, the first few scientifically trained whalewatchers were followed by travellers in their hundreds from all over the world. These people wanted to do what the locals had been doing for years: watch the blackfish. And furthermore, these travellers were willing to pay for the privilege. (The blackfish, meanwhile, had metamorphosed into orcas, their Latin name.)

These weren't the first tourists to discover the area. After World War I, when Canadian Pacific advertised that it "spanned the world," CPR steamships, the precursors of today's cruise ships, sailed up the Inside Passage to Alaska. Just as the CPR "sold" the Rockies to wealthy travellers, it "sold" the Inside Passage. All summer long these ships, immaculate in their black, white and buff company colours, appeared in even the smallest coastal communities, their rails lined with elegantly dressed travellers. At each stop the passengers went ashore and walked the length of the community, scrutinizing the inhabitants (even peering uninvited into baby carriages) much as they would have inspected the animals in a zoo. Then they returned to their ship, continued their voyage and the next week another group repeated the process. They didn't buy meals or accommodation in the communities they visited and, aside from merchant king John Wanamaker, who bought the entire front wall of a chief's longhouse and had it shipped back to New York, they didn't buy many souvenirs. They didn't make much contribution to the local economy. But their presence did introduce the coastal residents to tourists, a species that seemed just as inexplicable to them as the locals appeared to the visitors. Then World War II put an end to all this and, until the Island Highway was completed and the whales caught the public's attention, the area was virtually ignored.

Now a new wave of tourists had arrived. Unlike the wealthy and elite of the 1920s and '30s, these were usually middle-class people who had watched enough television not to expect First Nations people (for the Indian people had changed their name, too) to appear in feathered headdresses. These tourists arrived by car and bus, bought meals and accommodation, rented kayaks, bought souvenirs and frequented the beer parlours that had been spiffed up and now called themselves pubs. (Everything and everybody, it appeared, had undergone a change of name.) And what these people wanted to see, more than anything else, were orcas.

Of course, not being paid performers, the whales didn't always come forth on cue. They appeared seasonally and, even in season, they had their own agendas. Sometimes they were investigating one or another of the labyrinth of waterways to the north, or perhaps coursing up Johnstone Strait at a leisurely three knots en route to some predetermined destination. But that September day, in the bright sunshine, they were performing at Robson Bight. And there were two tour boats there. David surveyed them through an old pair of binoculars that were dented with dog bites. "That's the *Gikumi* up ahead. Can't see the name of the other one. Want to take a look?" The boats were perhaps 20 metres long and their decks were packed with people aiming video cameras and binoculars.

Orcas appear to have been designed by a graphic artist with an eye for a sleek, clean, modern design worked out in gloss black and white. Each time they hurtled up out of the sea—streaming water, vibrating with power—there was a collective "ah" from the boats and sometimes small shrieks that carried across the water.

"Go down and kill the power," said David. "Just shut it down and flick the battery switch over 'cause I had it on 18 amps charging the batteries."

Drifting, the *May S.* watched the whales and watched the tourists watching the whales. One of the orcas slipped below the surface on their starboard side, swam under them and surfaced on the other side just abeam the galley door. There was a great hollow whoosh and putrid whale breath wafted over the *May S.* "That breath sort of reminds me of one of your girlfriends," said David. "What was her name?"

Barrie ignored him. "Here we are trying to keep our distance," he said, "and they come right up beside us."

"I'm just too old-fashioned, I guess," said David. "I just couldn't see it. People are even out here lookin' at them in kayaks."

In an earlier era the Native people had plied these waters in their high-prowed Kwakwaka'wakw canoes, aware that below the water's surface lived the powerful max'inux, a fish hunter like themselves. But as technology arrived the Native people embraced it, building bigger and bigger boats with more power and more electronics. Fish became something you watched on sonar, and the immediacy of nature was removed. And then, suddenly, people who worked in office towers for most of the year were taking their holidays and sallying forth in kayaks, the frailest of craft, to experience nature as intimately as the Native people had done hundreds of years before. Plus ça change...

"Those kayakers have no idea," said David. "They think it's all fun. You're fishin' there. You've got your net out, eh. And all of a sudden 20 kayakers turn up. They all line up and go under your beach line. There's nothin' you can do. You know how dangerous it is with a beach line. It's killed lots of people."

Abruptly he restarted the engine, put the clutch in and eased the throttle up. "Let's go over to Double Bay and see how Bobby Barr's doing," he said. "I fished a lot with Bobby."

Double Bay used to be "the Buck's place." Del Buck was an electrician at the mill in Englewood until it closed. Nobody had heard of severance pay in 1941, so when Del's job disappeared he went across the strait and on Hanson Island found a piece of property with good moorage and fresh water. He built a house, a float, a boat, a ways and a three-metre fence to keep the deer out of his garden. Then he got a job as a fisheries warden and picked up a few extra dollars doing boat repairs. It was a way of life that required an astonishing degree of self-sufficiency, but it wasn't an unusual life for the times. The Buck's neighbours lived in much the same manner. They, too, inhabited similar little clearings lost in the convoluted folds of this immense wilderness, cut off from each other by an impenetrable forest and an icy sea. And they, too, had to venture out in their boats in order to buy groceries, kerosene, gasoline, and building supplies—or to see a doctor—on a stretch of water always swept by fierce tides and often roaring with southeasters.

The Bucks had been gone for years. Now at the entrance to the outer bay, propped in front of the rich green of the forest, there was a large, heavily braced plywood sign that read "Welcome to Double Bay Resort—Licensed Restaurant, Moorage, Bait." Neither the welcome nor the moorage was available to people like Bobby Barr. To circumvent this minor inconvenience, Bobby and his friends had anchored an old float out in the bay and set up a campsite that supplied the facilities their small boats lacked. There was a table made from driftwood planks and a motley collection of plastic chairs and small rusty barbecues. Scattered over the rest of the float were plastic buckets, piles of net, various bits of rope and assorted planks. When the *May S.* came around the point, Bobby was sitting there on an upturned milk crate, worrying away on a toothpick.

David brought his boat alongside, put her engine into reverse for a precise few seconds and she came to a stop with her hull a cigarette paper away from the float. Bobby shifted a pot on the Coleman stove and came over to take their lines and snub them around a loose board. David rested his forearms on the *May S.*'s dodger and looked down. "How you fellows doin'?" he said.

"Doin' fine. Fishin'. Fishin' rockfish. No salmon around," said Bobby. "Not like it used to be, eh, David? Got this boat and I sell to the Chinese. They like rockfish."

"I was talking to a fisheries guy the other day," David said, "and he said there's somethin' happenin' to the rock cod. They just about closed it down. They're tryin' to get everybody cut off on the codfish."

"Well, I'm tellin' ya', that's the only thing we're making money on now," said Bobby. "This last summer we were gettin' six dollars a pound for live rock cod."

Around them the shoreline's vivid green trees were perfectly reflected in the glassy water of the little bay.

"I read in the newspapers somewhere it takes 100 years for a rock cod to grow," David said.

"Maybe it can live to be 100," Barrie said, "but you can bet not many do."

Bobby's boat, the *Thumper*, was perhaps eight metres long, a double-ended dory burdened with a large homemade cabin. Her builder had made extensive use of marine-grade plywood and, unlike the *May S.*'s builder, hadn't been overly concerned with design. The cabin had very little headroom and not a lot of space. On the stern there were two rod holders for fishing rods and a cooler of sea water to keep his catch alive.

"When the rest of the guys come in we're goin' to have potluck supper," said Bobby. "We'll come over when we're finished. Where're you heading?"

"Goin' to town," said David.

"I'll bring you some fish to take to the family."

A diet rich in Coke, french fries and doughnuts had taken many of Bobby's teeth and left him with a layer of padding that strained the seams of his shirt. But neither fat, sugar nor hours spent sitting in the pouring rain jigging for cod had destroyed his good humour. He arrived and heaved himself over the *May S.*'s gunwale. His four friends, who, like him, spent their days smoking cigarettes, sitting still and eating junk, all looked like candidates for imminent cardiac arrest, but they, too, were smiling and carrying two huge red cod and an armload of smaller fish.

"No, no, that's too many," protested David.

But Bobby had the freezer open and was piling them in. "They're for the family," he said, shoving the last one in and snapping the lid shut.

They sat on the hatch and smoked and batted mosquitoes. After awhile one of Bobby's group said, "I'm gettin' a lot of smoke in the exhaust."

"You better grind your valves," said Barrie.

"Well, I'm hopin' it's the valves. I'm hopin' it's not the rings. Because I got no money if it's the rings."

"Never mind we didn't save the salmon," Barrie said. "We didn't even save the money."

"Boy, you talk about money," Bobby said. "I went down to Seattle once, after fishin'. Pockets just full of money. Saw a speedboat there at some yacht club. Beautiful little boat. About 35 foot. One of them Chris-Crafts. I asked the guy that owned it what he wanted to take me for a ride." Bobby squinted through the smoke from his cigarette. Even now he could remember the growling twin Chryslers and the way his head snapped back as they took off and tore across Lake Union. "Lotsa power," he said.

For awhile no one spoke. It was mercury vapour lights that got them talking again.

"We came through the Hole in the Wall," David said. "It was eight o'clock at night. We just used the light. It lit up the whole place like daylight. We had it on up by the Yucultas. We were just about up to Powell River and a boat phoned and said, 'I'm six miles away and I can't see anything. Will you turn off that light?'"

"You're not supposed to run with those lights on."

"What I'm saying is, there's no traffic out there. We ran all the way from the Yucultas to Powell River and didn't pass a boat. That tells you how much traffic is on this coast now."

After a pause Barrie said, "Are you going to fish here all winter, Bobby? Not much protection in that little boat in the rain."

"Oh, we've got the gillnetter. We can sit in there at night and watch TV and get warmed up. Yep, this is my permanent address." He chuckled and rummaged around for his wallet. "Here's my new driver's licence. Had to have some ID to cash my cheques. The girl wanted my address. I told her I don't have one; I just stay on my boat at Double Bay. But she says you've got to have an address."

"Barr, Robert," it said on the licence. Down at the bottom was his address: "Wharf Street, Double Bay, BC."

Later, when David and Barrie had rolled into their sleeping bags, Bobby and his friends were still out there, sitting on the milk crates in the darkness laughing away about Lord knows what.

CHAPTER 2

High Boats:
Gospack — Harry Stauffer (BCP)
Nasoga — Charles Wilson (ABC)
Kitgora — Steve Warren (Can. Fish)

Pioneer Journal, July 22, 1953

Barrie gave the frying pan a final swipe and slid it under the little cookstove. "We on our way to town, then?"

"How long have you got? I mean how long have you got 'til Linda sends out a search party for you?" said David.

"What are you getting at? You want to go back to the Bay and hang a net and try to get a few fish?"

"No," said David. "I know it sounds kinda dumb but ever since my dad passed away I've been thinkin' of all the places he took us to when we were young. I just thought if we got the weather and if you didn't have to get back to town real quick we could go to a few places. When I'm fishin' I'm too gung ho. Just never take the time."

"I've never forgotten those places your dad took us—and the places I got to go on the *Kitgora* with Steve. Always wanted to go back."

And so they slipped out into Blackfish Sound in the fog and headed for Mamalilaculla.

In August and early September there is always a bank of fog lying over against the mainland. At night it rolls west to Vancouver Island. By day it gradually disperses in the west, but in the east, below the Coast Mountains, it remains—a curtain of grey that blots out the horizon. Now at 10 in the morning there wasn't enough heat in the sun to burn off the mist that clung to the water of the sound, but as the air warmed it began to drift and thin. Here and there, as it ebbed and flowed, tiny grey granite islands tufted with brilliant green trees rose out of the pale vapour. In places where the sun broke through, the mist fell away revealing an ocean spitting light from every cat's-paw wavelet.

Barrie stared down at the tide racing over the surface of the water like liquid pouring over a sheet of glass.

"Hard to believe people used to paddle across here in a hollowed-out log," he said.

"It was a big log," David grinned.

"But I'd rather do it on the *May* and so would you."

The *May S.*, her engine throbbing evenly, threaded her way through a maze of small islands. On each a flurry of salal and salmonberry and a few shaggy cedars were arranged with asymmetrical taste. On some, the rocky shore gave way to tiny beaches.

David said, "With the fog and all you can't see the clear-cuts. Out here right now it probably looks just the way it did a coupla thousand years ago."

"That's when your great-great-great-auntie paddled across here all the time."

"Not *my* great-auntie," said David. "My great-grandfather was an American and my great-grandmother was a Tlingit woman."

None of David's forebears, then, were anywhere near the area when Captain Vancouver's ship HMS *Discovery* reached the mouth of what is now the Nimpkish River. Vancouver arrived shortly after 10 p.m. on July 19, 1792, and anchored just outside a stretch of sand and sedge grass that formed a tiny island there. At that latitude, that time of year and that hour, there was just enough light left to distinguish the mouth of the river and, across the strait, the long mound of an island. And there would have been just enough light for the Native inhabitants of the village at the mouth of the river to witness the arrival of the *Discovery*.

As impressive as the ship's appearance must have been to a people who travelled in nothing larger than a canoe, it probably wasn't the first time they had seen a sailing ship. In the last quarter of the 18th century Spanish, British and French explorers had found their way to the coast of what was to become British Columbia. In their unwieldy sailing vessels they fumbled through its narrow channels, ferocious tides, rock-choked passages, its reefs and its fog. They navigated without the aid of GPS, the sweeping line of a radar or the flashing numbers of a depth sounder. They didn't even have a chart (for in time the unwieldy ships' purpose was to survey this coast and *produce* a chart).

Lacking technology, these foreign explorers used, as a substitute, seamanship—and local knowledge. For this wasn't a totally empty wilderness, however much it might appear to be. Its inhabitants were scattered bands of Native people who lived in villages lost in its vast jigsaw puzzle of land and sea. They were a people forced to the water's edge by immense trees and a jungle of undergrowth. Because their sustenance came from the sea, they had become skilled seamen and canoe builders, wise in the ways of this treacherous coast and its complicated tides. Consulted by the officers of their various majesties' ships, they gave advice, warned of hazards and shared a hard-won knowledge.

The village at the mouth of the Nimpkish River was situated just where the river's channel broadens and joins the sea. On the south side the river spreads over low land, creating patterns of tiny rivulets winding around small islands of sedge. But on the north side of the Nimpkish, the water is contained by a high steep bank. This feature gave the settlement its original name: *Whulk*, or bluff. Still visible on the slope of that high bank is a rectangle of grass as big as a football field. On the central coast there are no fields—except those hacked out of the rain forest by man—and this one is all that remains of the village of Cheslakee.

To clear this hillside with nothing but primitive tools and fire must have been a Herculean task, but the field existed 100 years ago and who knows how many hundreds of years before that. In 1792 there were 34 houses there and 500 people. It's this grassy hill with its rows of dwellings that is pictured in the engraving copied from a sketch by John Sykes, one of Captain Vancouver's midshipmen.

It was another 50 years before ships of the British Royal Navy began to map the area. Her Majesty's ships *Alert* and *Cormorant* were the first to undertake this task. By 1857 the steam engine had come into existence, so that when the 440-tonne survey ship HMS *Plumper* took up her part of the detailed mapping of coastal waters (an endeavour that took 13 years), she had an engine that could drive her along at 6.5 knots. By 1860 Johnstone and Broughton straits, Queen Charlotte Sound, Knight Inlet and "adjacent channels" had been surveyed and mapped in the London offices of the British Hydrographic Service. For three shillings the Royal Navy sold a chart that could be used with confidence today. The Native village at the mouth of the Nimpkish River was identified as Cheslakee after its chief, and the sandy islet became Flagstaff Island. The island across the strait was named after the *Cormorant* and the curve of bay after the *Alert*.

The incessant rain that falls on the north coast grows trees better than anything else. Aside from the huckleberries, salmonberries and salal berries that grow in the dense undergrowth, it was the sea that provided the residents of Cheslakee with their food. In the clear frigid water five species of salmon spawned in such numbers that old-timers remember a solid mass of fish. There were cod, halibut and red snapper out in the straits, eulachon at Knight Inlet, and crabs, mussels, abalone, prawns and clams. Catching, cleaning and preserving enough of this abundance to ensure a year-long food supply took the combined efforts of everyone in the village, including a few unfortunate slaves. Just acquiring the necessary fishing gear required the many hours of work that substituted for money. The Native people made fishing line and nets from inner cedar bark, nettle fibre and kelp. They fashioned hooks from bent wood and splinters of bone, and made spears and herring rakes. They built traps and weirs and stone dams on the river, and trolled from their canoes. On the shore the grandmothers tended beach fires ringed with racks of fish impaled on saplings. A crowd of ragged children kept them supplied with driftwood.

Nobody took the harvest for granted. Pounded into their collective consciousness was an awareness of the power and brutality and fickleness of nature. When it was too rough to fish from a canoe, or when the salmon runs were sparse, or when it was too wet to smoke the fish properly, the result was privation if not starvation. To avoid such catastrophes these people tiptoed through a world populated by powerful, easily offended spirits. They tiptoed at their own pace, however, for an absence of pressure was the one luxury that their mode of existence provided.

It's not uncommon to hear contemporary individuals romanticizing the aboriginal way of life beyond all recognition. Filtered through time and European sensibilities, their version of those early Native cultures bears little resemblance to the real thing. Captain Prevost of HMS *Virago* was there at the time and his impressions were more accurate. He was haunted by "the depravity and deprivation of the Indians of the Northwest," and when he returned to England he appealed to the Church Missionary Society for help.

The society responded with money and missionaries. These latter were people who practised Christianity in its most basic and practical manner. They travelled halfway across the world to this isolated coast, and in due time those who settled the North Island established a farm on Cormorant Island, built a small tuberculosis sanitarium to care for the many Native people with this disease and offered schooling to provide literacy and some basic skills. They were selfless, devoted people who gave their lives to what they believed to be a pressing need. In their eyes, "improving" the Native peoples' lives (and they genuinely wanted to improve their lives) involved changing them completely. What a Victorian Englishman or Englishwoman was incapable of recognizing was that one couldn't—and shouldn't—discard one culture for another as one would change an overcoat.

Long before the missionaries or the map-makers, however, white men had appeared on the coast doing what white men seem compelled to do: to look for business opportunities. The Russian and

American fur traders were the first. The fur trader Joseph Ingraham cruised the Queen Charlottes in 1791, but there had been others well before him and there were others long after.

By December of 1848, however, the search for opportunity was centred farther south. Word was out that there were sizable chunks of gold in the rivers of northern California, and the rush was on. A comparable situation couldn't occur today; it's difficult to imagine what it must have been like to wake up to the news that all and sundry have a chance to wrest a fortune from the wilderness. The only prerequisites were a strong constitution, a sense of adventure and a healthy dose of greed. Success wasn't guaranteed, but it had been achieved by a sufficient number of gold seekers to provide a surging motivation for the rest. It was a scenario guaranteed to attract not only the venturesome but also the unsavoury, who circled the horde of gold seekers like sharks, picking off the unwary before they even reached the goldfields.

All the above provided an irresistible attraction for a young man of 26 living in New York, the city of his birth. His name was Alden Westly Huson. The Huson family were United Empire Loyalists, descended, it has been supposed, from Sir John Temling Huson (or Houghson). "Wes" Huson's great-grandfather was Thomas Huson, a general in the British army killed at the Battle of White Plains. Like his great-grandfather, Wes was genetically disposed toward action and adventure, and word of the gold strikes in California promised him just that. He left New York, travelled south to the isthmus of Panama, crossed it by rail and journeyed up the Pacific coast to San Francisco. Perhaps he arrived in California too late to get in on the bonanza; perhaps all the best claims were already spoken for. In any case, having reached his destination he didn't linger in the goldfields of California or in the diggings in the Columbia River area. Instead he hurried north to Canada. He arrived in Victoria in 1858 as one of a horde of gold seekers hastening toward the Cariboo, and he found himself in a little settlement overwhelmed by the influx. Harry Gregson's *A History of Victoria* describes the scene:

> The townspeople were just leaving church to return to their whitewashed cottages on Sunday morning, April 25, 1858 when the Commodore, a wooden sidewheel American steamer brought the first shipload of gold-seekers. From their position on the [Christ Church] hillside they watched fascinated as she disembarked a stream of men, most of them wearing red flannel shirts and carrying packs containing blankets, miners' washpans, spades and firearms. The first large "non-British" element had arrived in the colony.
>
> Of the 450 men in the party only 60 were British subjects. Some had money, some were penniless. Most carried bowie knives and revolvers.
>
> Victoria with its few hundred settled white residents, its picnics, intimate parties and amateur theatricals, its politicking and Douglas-baiting had to cater to what some authorities estimated at 25,000 gold seekers.

Amazingly enough, it did cater. A tent city sprang up and brought with it all the problems that we now associate with refugee camps. Water, food and sanitation were in short supply; liquor was not. It was James Douglas's firm hand and his experience as chief factor for the Hudson's Bay Company that brought order out of this chaos. And James Douglas intended that Victoria should not only survive this invasion but profit from it. Under his "good commonsense," and with clout provided by Her Majesty's warships, he kept order and laid the groundwork for what was to become a prosperous little city.

In 1860 it still had a long way to go on the road to civilization, however. An account of an examination conducted by a justice of the peace on June 25 of that year portrays a Victoria that is closer to Dodge City than Devonshire. In this case of assault with intent to cause grievous bodily harm, Alden Westly Huson gave evidence as follows:

> "About eleven o'clock last night I told my woman to take a candle and go down to her aunt's house to see if

> her aunt wanted anything, as the Kanaka, Palow, had been beating her in the afternoon. She went down and left the door of the house open and a few minutes after I went in myself. When I stepped into the house there was no one there besides the two women and they were talking in their own language. I immediately closed the door and examined the lock which had been broken in the afternoon. Soon after one of the women said she heard the Kanaka's voice again coming from the street. I stood back about three feet from the door; when the Kanaka burst the door open and raised his arm to strike a blow at one of the women I endeavoured to ward off the blow and received a wound in the arm. The Kanaka had jumped to the middle of the floor and was within reach of the woman. I distinctly saw the knife just as I received the wound. My arm immediately became powerless and I struck at him with my other hand at the same time he struck the woman Mary."

Hearing Huson's shouts for assistance, one George Collins entered the fray. "When I got there," he said, "Mr. Huson and the Kanaka were scuffling and the latter had a knife in his hand. He struck at me and I jumped back and struck him with a stick and knocked him down."

The Wild West, it appeared, was just as full of action-packed adventure as Wes Huson had imagined it to be, and he was equal to its challenges. He was prepared, as well, to turn his hand to anything that would give him a living; he could not afford to be overly fastidious in his choice. In October 1867 he was advertised in the *Daily British Colonist* as the proprietor of Victoria's notorious Adelphi Saloon. It wasn't long before he became dissatisfied with the role of saloon keeper, however. He had come to British Columbia to make nothing less than a fortune, and like all prospectors he was driven by dreams. So he left Victoria and the bawdy Adelphi and set out to explore the northern coast.

An early resident of Alert Bay, now long gone, remembered hearing that Wes had arrived in a sailing sloop. Perhaps he had travelled with one of the many traders who plied the coast with goods to sell to the Natives. Like the men who headed to the Fraser, Wes, too, was searching for a fortune, but instead of following the crowd he struck out on his own. And unlike the crowd, who were fixated on gold, he believed that there were other minerals out there that could also make a man wealthy.

At Alert Bay he came ashore and made this Kwakwaka'wakw seasonal encampment his home base while exploring the surrounding country. Contemporary accounts mention a lone white man living among several hundred aboriginals in the summer village on Cormorant Island. Prospecting was his focus; he felt certain that this vast wilderness held untold treasures if he could just find them. He must have been indefatigable, for he bushwhacked through the jungle-like undergrowth of the North Island from the east coast to the west and later was one of the first white men to cross the Coast Mountains to Bella Coola.

Then he came within a hair's breadth of becoming a coal baron. At Fort Rupert he was told of a coal deposit at nearby Suquash, and he promptly applied for a lease of the property. In March 1868 he presented William Trutch, the surveyor general in Victoria, with a detailed proposal for the development of the North Pacific Coal Company. His application was signed by himself, his father (who had joined him by this time) and a Mr. G.H. Fish—this latter person obviously a Canadian. For Wes Huson had an added obstacle to overcome in his search for a business opportunity. As an American he was prevented from taking part in many business transactions unless a Canadian was involved. Hence Mr. Fish. In any case the North Pacific Coal Company received its lease. At the bottom of an internal government memo is the following notation: "The applicants are, I believe, practical men, but I fear that they do not possess any means. I think, however, every encouragement should be given..."

Means or not, Wes Huson worked hard to establish the mine. If it hadn't been for Robert Dunsmuir, who found better-quality coal much nearer the markets, Wes Huson might have been the one to

build himself a castle in Victoria. Instead, when the anthracite coal found near Nanaimo destroyed any hopes for the mine at Suquash, he continued his search for minerals. Periodically he boarded the SS *Beaver* and turned up in Victoria with canvas bags filled with samples of copper and iron ore and Haddington Island granite. This latter was the material eventually used to build the Parliament Buildings in Victoria, and Huson helped develop the quarry that resulted. All the minerals that he found did, indeed, exist in viable quantities. A hundred years later Empire Development and Island Copper Mines gnawed vast holes in the wilderness and filled ore ships that sailed away to Japan. But in the 1860s and '70s Wes Huson was ahead of his time—and perhaps short of capital—and none of these discoveries made him the fortune he was seeking.

In 1873, when Huson was 41, he married Mary Ekegat, the Tlingit woman with whom he had been living. Mary was born in Alaska. To escape an early and abusive marriage she had fled south to live with the Hunt family in Fort Rupert, and it was here that Wes Huson met her. The marriage certificate notes her age at "about 25 years." A studio portrait taken about this time portrays a woman with real presence. Handsome and dignified, she is dressed in sober Victorian finery; her level gaze compels your attention. Her husband's photograph is that of a good-looking greying man with a short, neatly trimmed beard. Perhaps the person who took these two portraits was Stephen Spencer, the Victoria photographer who became Wes Huson's friend and shared his entrepreneurial bent. By the 1870s these two were considering a way of making money that had nothing to do with prospecting.

The mild curing of salmon in commercial quantities had only recently been perfected, and on the southern BC coast there were commercial salteries processing salmon this way. Then the first cannery was

Opposite: A portrait of a handsome, neatly turned-out Alden Westly Huson, David Huson's great-grandfather, circa 1873, the year the 41-year-old married Mary Ekegat. From the collection of David Huson

established on the lower Fraser River. By 1871 canned, salted and pickled salmon was being produced and exported in large quantities. Spencer and Huson pooled their slender resources and built a primitive little saltery on the waterfront in Alert Bay, where there was deepwater moorage for boats and easy access to the huge salmon runs in the nearby Nimpkish River.

As a location, Alert Bay had an incomparable advantage. Not only were the Nimpkish salmon plentiful but their firm texture proved ideal for salting and made them superior to salmon caught elsewhere. Even so, Stephen Spencer was having little success flogging the product in Victoria. A letter to "Friend West" dated 1876 says, "There doesn't seem to be any demand for the fish at present. I have tried the Chinamen as well as the white market and finally I have left them with Earle to try and sell them in his store to try and get them started. Dr. Tolmie got his fish—the other barrel I have on hand. The smoked salmon were first class. I gave some of them around to let them see what you could do and they all thought they were first quality."

"They" didn't buy them, though; the saltery's only customers were the Japanese. These people, discriminating consumers from the very beginning, had discovered that the salmon from the Nimpkish River and Simoom Sound were much firmer in texture than those from other areas. By 1878 a Mr. Fujiyama and his countryman, Mr. Sukiyama, knowledgeable in their customers' preferences, were in charge of the saltery that Spencer and Huson had built. Before long steamships were calling to take the 1.5-metre wooden boxes of fish to Vancouver and then to Japan. A year later Wes Huson, eager to tie down his legal rights to this fish supply, advertised in the *Daily British Colonist* his intention to purchase 65 hectares at the mouth of the Nimpkish River from the Crown.

It wasn't the ownership of the land that really changed anything,

OPPOSITE: Mary Lyons Huson, nee Ekegat, was 'about 25 years' of age when she married Wes Huson and posed for this portrait. Her level gaze may reflect the self-reliance learned at an even younger age, when she fled an abusive husband in Alaska. From the collection of David Huson

however. It was the introduction of commerce. The saltery had secured a supply of fish but had no one to process it. In this sparsely populated wilderness, the only available labourers were the peripatetic Native people who moved back and forth from Cheslakee village—and from much farther afield if there was a wedding or a potlatch to be attended. The book *Kwakwaka'wakw Settlements 1775-1920* states, "There is conflicting evidence about when the inhabitants of Whulk crossed to Alert Bay. Part of the problem is that Whulk continued to be a fishing site after it was abandoned as a winter village."

Persuading these inhabitants, whether from Whulk or Alert Bay, to come and work at the saltery on a regular basis was more difficult than Huson and Spencer expected. For they found themselves involved in the introduction of a radical idea—the idea of "working for a living"—to a people for whom that concept was entirely foreign. The aboriginal population worked only long enough to exist on a subsistence level; their days weren't "workdays" as such but days in which work, play, art, music and celebration all blended seamlessly into a whole. Untouched by the Puritan work ethic, the men found nothing shameful about spending whole days talking or gambling or simply sitting in the sun doing nothing at all. These people paddled across to Alert Bay and turned up at the cannery out of curiosity or when the spirit moved them, but the concept of a regular workday had no appeal.

It took a combination of two divergent values—spiritualism and materialism—to solve the problem. First Wes Huson persuaded Rev. Alfred Hall, who had set up a mission at Fort Rupert, to move to Alert Bay. He pointed out that Alert Bay was a more central position in relation to the surrounding villages, and he promised to supply the mission with land and build a mission house. Thus convinced that the move would be an advantageous one, Rev. and Mrs. Hall moved to Alert Bay in 1878, the same year the saltery was established.

The other lure took the form of material goods. There was a store in Alert Bay now. Periodically a steamer called in with supplies and goods of all kinds, which could be bought with wages. This proved an irresistible attraction. Alert Bay might owe its name to the survey ship HMS *Alert* but it owed its existence to people like Stephen Spencer and

Persuaded by Wes Huson that Alert Bay would be a more central location for their mission, Reverend and Mrs. Alfred Hall moved there from Fort Rupert. Photo courtesy BC Archives, E-09190

Wes Huson. Capitalism had arrived and the Native way of life was changed forever.

By the late 1800s a footpath followed the pebbly beach that curved along Cormorant Island's wide bay. It stretched a kilometre or two from the mission house and the little sawmill at one end to early settler John Robilliard's log house at the other. In between lay the First Nation village, marked by a row of massive longhouses standing shoulder to shoulder, their great flat facades facing the sea. The distinctive shallow-pitched rooflines proclaimed the Kwakwaka'wakw cultural heritage, as did the totem poles that towered in front, the fierce faces of wolves and bears and thunderbirds glaring down at passersby. The only break with tradition was the milled siding that covered the log frames in place of long split-cedar shakes. Here and there a Native inhabitant of the village had opted for the more "modern" style of a conventional frame building. Here and there the gigantic log frame of a longhouse under construction raised its bulk. For this was a community still coming into existence.

Lying in front of the longhouses or pulled up on the beach below were the occupant's high-prowed dugout canoes. And all along the waterfront was a raised platform built out over the beach. This was the Kwakwaka'wakw equivalent of the Englishman's club and its philosophy was *carpe diem*. Here was where the men indulged their passion for gambling games and carried on interminable conversations.

The beach itself was the domain of the grandmothers. As they did on the beach at the mouth of the Nimpkish River, they built big fires that burned continuously in the fishing season. The fires were encircled by an intricate network of saplings and filleted fish and, as at Cheslakee, a gaggle of barefoot boys kept them supplied with driftwood.

Establishing this village within a village had caused some difficulty. Huson had originally obtained a lease for the whole of Cormorant Island. Now some of this land had to be extracted from the lease agreement and designated as a reserve for the people who had lived there first.

A spacious mission house was sited on a portion of this land, and when it was completed the Halls left the wilds of Fort Rupert for the dubious charms of Alert Bay. They came to a bleak little settlement that would have been insupportable for a depressive. The incessant rain bleached the sea, the beach, the piles of driftwood, the handful of frame buildings and the row of longhouses to a weathered grey. Even the dark trees were greyed with mist. Then the impenetrable darkness of a wilderness night obliterated it all, and there were only squares of light from windows gently lit with kerosene lamps and a bobbing light from a lantern carried by someone walking along the waterside path.

The Halls were strengthened with purpose, however, and they wasted no time in putting their new accommodation to use. They started a school for the aboriginal children, and the lumber that sheathed the new longhouses came from a sawmill Rev. Hall established to provide the boys with a trade.

The Halls were soon recognized as the support of anyone in need; two of those in need were a boy and a girl, both of whom had Native mothers and white fathers. Stephen Cook's father was an English shipbuilder and Stephen was born in Victoria. His mother, a member of a high-ranking family, married a second time to one of her own people. But she was widowed shortly afterward and returned with her father to Alert Bay, where she brought her son to the Halls to be educated. Jane's mother was a chief's daughter from Fort Rupert and her father an English sea captain. At 14, she too came to live with the Halls and receive an education.

Both children were bright and capable and were reared to respect Christian principles. Rev. Hall looked forward to the time when Stephen would himself become a missionary. It was a visiting English professor who thought otherwise. How he came to be there is immaterial; it was his observation that changed Stephen's direction in life. "The boy's not fitted for your work," he said. "He should go into business." Rev. Hall evidently respected his guest's judgment for he abandoned his plans for Stephen and instead made him his secretary and bookkeeper and finally the foreman of the mill.

It was soon apparent that the English professor was right. Stephen was a devout Christian—and a born entrepreneur. Working in the sawmill he learned from the Japanese, who were in the area by now, that yellow cedar was in great demand in Japan for boat lumber. Stephen Cook knew where the yellow cedar grew. In partnership with Spencer and Huson (the latter, no doubt, always happy to have the required Canadian associate), he staked out a timber limit. Steve Warren, Stephen Cook's grandson, said, "They pounded the stakes in one day and the next day they sold the timber to the Japanese." The partners divided the money three ways and Stephen used his share of the profits to build a dock and a store.

The Kwakwaka'wakw people, fishermen par excellence, were by now a vital part of this new enterprise. Paddling across the strait to the mouth of the Nimpkish, they trolled the tides from their dugout canoes and then delivered their catch directly to the saltery, where the women cleaned, salted and packed it. By 1881 the saltery's owners had become convinced that canned salmon was the product of the future, so Wes Huson bought canning equipment and converted the Alert Bay saltery to a cannery. The little mission sawmill, originally intended to teach skills and supply the community with lumber, became a bona fide box factory, and the man-and-wife trollers now caught the first Nimpkish River salmon to be put in tins. In Victoria Stephen Spencer continued to do his best to market this new product. He was not particularly successful.

In 1901 the Rt. Rev. W.W. Perrin, Bishop of the Diocese of BC, visited Alert Bay as part of his duties. This man was one of an astounding group of clergy that fanned out all across the world promoting the religion of Victorian England, an era that produced a society that was prim, narrow-minded and utterly convinced of its superiority. As a consequence, the Right Reverend was insensitive to the fact that a mission like his was ill-advised. Instead, eager as ever to minister to "the heathen," he disembarked from one of the Union steamships and plunged into a busy round of activities.

He was distressed, he noted later, by the First Nation marriage customs. "The whole question of their marriage customs is full of difficulties. A girl is sold at a very young age to her husband but as soon as she has paid back the purchase-money, she is free to leave her husband without disgrace and to be married to another who may be willing to give a larger price for her. In this way a young woman of 21 may have lived with four different men and the result is disastrous."

(From today's vantage point this arrangement may seem an enlightened response to the problems of monogamy, but to a Victorian Englishman it certainly didn't.)

On the other hand the Rt. Rev. Perrin was considerably buoyed up by his visit to the school. "The school children are quite equal to any white children in secular knowledge," he wrote, "and I only wish that other school children in Canada and England had an equal knowledge of their Bibles."

And he was concerned, as many after him were, about the liquor problem. But all in all it had been, he thought, an encouraging visit. He commended Rev. and Mrs. Hall for their devoted service and reboarded the steamship on its return trip to Vancouver, his heart "overflowing with thankfulness for all His mercies." The bishop's visit served, among other things, to underline the fact that the residents of Alert Bay represented a broad spectrum of humanity.

Wes Huson, one of those residents, was again struggling to keep a fledgling business afloat. The government official who had approved his lease for the Suquash coal property noted that Huson and his partners "possessed little means." Lack of capital was certainly one of his problems; another was lack of business contacts. But perhaps his biggest obstacle was his predilection for prospecting. Stephen Spencer was having great difficulty interesting people in Victoria in canned salmon, and it seemed to a discouraged Huson that canning fish was nothing more than a sideline distracting him from the possibilities that prospecting offered. In 1884, just three years after the saltery became a cannery, Wes Huson sold his lease on the 250 hectares on Cormorant Island that he had obtained in 1870 and left the fish business for good. Stephen Spencer and his new partner, Thomas Earle, bought the lease for $1,000. In Earle, Stephen Spencer had finally found what was needed: a partner with capital, business experience and invaluable contacts in London.

Wes Huson and his wife were to have 10 children, but in the 19th century a child's chance of reaching adulthood was far from assured. In infancy and early childhood five of the Huson children succumbed to illnesses. And in 1892 Mary herself died. She was only 44. The older Huson sons scattered, picking up work where they could find it—in the Yukon, on the Skeena River, crewing on ships, logging on Swanson Island. Always they kept in touch with their father; always they kept

Bighouses crowd the waterfront, the Alert Bay version of Main Street. Photo courtesy UBC Library Rare Books & Special Collections, Fisherman Publishing Society

Salmon from Wes Huson's cannery made its way to market under a variety of brand names. From the collection of Barrie McClung

their eyes open for prospecting opportunities; and occasionally they helped Wes with his ventures. A younger son, Spencer, named for his father's friend and former partner, made *his* contribution by becoming a first-rate hunter. He kept the family supplied with deer, ducks and geese and sold the excess. Like most households of the time, theirs had a hand-to-mouth existence; unlike most households, however, they lived in hope. Until blindness made it impossible, Wes Huson continued to file mining claims and collect ore samples, convinced that one of them would prove to be the motherlode.

In the census of 1881 Wes Huson's profession was listed as "trader," but when he died at Alert Bay on December 19, 1912, the Record of Burials listed his profession as "canner," which was perhaps more appropriate. For although Wes Huson never did make the fortune that seemed so attainable to him as a young man in his 20s, the little saltery he established with his partner Stephen Spencer changed the lives of the inhabitants of Alert Bay forever.

CHAPTER 3

High Boats:
Chief Takush — Simon Beans (BCP)
White Wave — John Bell (ABC)

PIONEER JOURNAL, SEPTEMBER 10, 1952

For kilometres of convoluted coastline, down channels, around points, past islands, there are only trees—as thick and bright as grass—right down to the metre or so of grey rocks that rise above the high water mark. To the uninitiated one route looks much like another, but the *May S.*, sure of her direction, threaded her way through the islands, leaving a lacy white wake behind her.

Up on the dodger David and Barrie watched a float plane, its engine flat out, struggle for more than a kilometre to get airborne. Obviously overloaded, it motored back to some invisible settlement where it must have discharged cargo, for presently it appeared around a point and tried again. It lifted off and zoomed along one metre off the water, three metres, then 20. At 200 it banked and flew southward toward Johnstone Strait. The *May S.* was alone again.

"Here," said David. "Junior man steers. Just hold her well off that point 'cause there's a reef comes right out about a thousand feet. I want to go below and check the coolin' water. We had a head gasket go last year and a bunch of oil got in it. We flushed it a few times but there's still some in it."

"If you jam a wad of foam in the filler neck and change it often," Barrie said, "the oil saturates the foam while the water tends to flow through it. It's surprising how much it takes out."

David was gone for a good half-hour. He arrived back with two very greasy hands, a banged-up knuckle with a rag wrapped around it and some rather Spartan ham sandwiches.

"Hope you like ham. We got no lettuce and I can't find the mustard. I guess those guys at the Bay that were lookin' after the boat cleaned me out of stuff."

"With no refrigeration on they probably dumped it over the side and did you a favour."

David picked up the binoculars and studied the landscape in front of them. "You remember how it used to be in here with all the boats? Now there's nothing left at Gilford Island village. Just old people and a few kids. Everyone just left, I guess."

"I was over visiting Steve," Barrie said, "and he was upset because they tore down the old cannery at Bones Bay because it was a hazard or a liability or something. I can remember the first time we went there with the old *Kitgora*. It was sure an experience for me—a kid from the city. And now they're talkin' about tearin' down the cannery at Alert Bay."

"They wouldn't tear that down, would they?"

"Seems like they might."

"Well, if they do it'll be a sad day for me. My grandfather, my father and me, we all bought our nets there and we hung our nets there. I can't see them tearin' it down. I mean, the roof needs fixin' in places. It wouldn't seem like the Bay with that gone."

"I don't see why they wouldn't make it into a tourist attraction. It's all tourists now. That's the future."

"Not a lot of money in it, though, is there? Gilbert takes people out on the *Cape Cook*. And they've got a bus that takes people down to the big house. The people do a couple of dances and sing a couple of songs, and they all go away—experts on our culture. But the dancers get paid a little so it's somethin' for them."

To the confusion of its inhabitants, Alert Bay's focus has reversed itself. The emphasis now is on marketing culture, whereas in the late 1800s, far from encouraging singing and dancing, commerce was trying to replace those activities with a regular workday—and the idea was slowly catching on.

Cannery buildings have dominated Alert Bay's waterfront for most of the town's life. Photo courtesy UBC Library Rare Books & Special Collections, Fisherman Publishing Society

In 1884 Stephen Spencer and his new partner, Thomas Earle, switched from salting salmon to canning it and shortly after that sold their flourishing business to G.I. Wilson and Henry Doyle. Between that time and its emergence as BC Packers Ltd. in 1928, it changed hands three times and each time its name was subtly different. In 1909 Wilson and Doyle sold the company to a man called Barker from Tacoma and it became the BC Packing & Fishing Co. It was sold again in 1914, and when it was reorganized in 1928 it assumed the name BC Packers. By 1908 the cannery, by whatever name, was leasing the mission sawmill. Then, in an effort to ensure its supply of fish, it established a hatchery on the Nimpkish River. An indication of the growing importance of the fishing industry was the presence, as early as 1909, of academics sent out by the government from some of the eastern universities to study the life cycle and habits of the sockeye.

This thriving industry required more fish than the trollers in their dugout canoes could provide, so they were gradually replaced by fishermen using nets. All along the coast the fishing companies were building their own fleets of boats—big wooden skiffs some eight metres long with space for three or four oarsmen or a sail. On their sterns lay mounds of net complete with cork and lead lines. On a favourable tide, a gas boat towed this cluster of skiffs to the fishing grounds and then towed the skiffs and their catch back to the cannery. At the Nimpkish River nets were set out to encompass the salmon schooling near the mouth of the river. A cork line kept the top of the net on the surface of the water and a lead line held the lower end on the bottom. As both ends of the net were pulled in they brought the fish with them. The fishermen were largely Native, but now there was the occasional white men among them, too.

Sam Hunt was 16 at the time. "Up Nimpkish River," he said, "they had these stakes driven in that you tied the end of your net to. You got there bright and early in the morning to be at the spot you wanted to be in. There were stakes for a mile up and down the river. So you'd go out in the dark and you'd row this skiff—six men rowing and throwing the net, and the skipper stood at the bow with an oar. And you'd row out. The fish were coming up the river and you're slowly going down as they're coming in. And then you'd come ashore and you'd pull the net on the beach. This other man, he drifted down with the end of the net until he come up to where the men were pulling the net ashore. They pull the fish onto the beach and then roll them into a skiff. And it was all by hand. We got 500 to 550 one set. That first set we paid all our riggin'—gumboots and everything. I went home with $180 to $200 for three weeks, but for a school kid that was pretty good going."

This was drag seining. Now, instead of catching the fish one at a time, these red spring salmon, the fish so prized by the early Japanese consumers, were caught in great slithering masses. The anthropologist Franz Boas was in Alert Bay in 1900, and in a letter to his children he recorded average catches of 1,000 salmon a day, with 5,000 caught one day he was there. In that same letter he described the canning process:

> This building stands on legs in the water, and the floor has half-inch spaces between the boards so that the blood and the intestines can easily be washed through. The fish are put into a heap at one end of the room.

Fishermen haul in—by hand—a drag-seine net teeming with salmon, circa 1918.
Photos from the collection of the author

Drag seining in the Nimpkish River. One end of a net is tied to a stake, and the rest is paid out as the boat is rowed across the river. A cork line keeps the top of the net floating, while a lead line keeps the lower edge of the net on the river bottom. Fish coming up the river are trapped, the boat circles back to the stake to close the net, and then it is hauled ashore.

> From the heap a man takes a fish and puts it on a table. A Chinese worker takes it and cuts off the head, the fins, and the tail with a sharp knife. First, however, he opens the belly and takes out the intestines. Then he throws the fish to the left to another Chinese who removes the scales and throws it into a large wooden rinsing tub where it is cleaned. Then they are cleaned again and given to an Indian who has a little machine consisting of four knives. The fish is put under the machine and cut into four parts.... The women then take the fish pieces and put them into the cans.... Another machine puts the tops on the cans.... Then about 144 cans are put on an iron frame.... Then the whole frame is picked up ... and put into a boiler.
>
> The heads and intestines are thrown into the sea and little dog sharks come and eat them. You should see the number of sharks around on days like this.

Because the Nimpkish salmon were such a prime product, some of the fish didn't go to the cannery. Instead they were cleaned, packed and loaded onto the packer *Chasm*, which raced them to the city at a good eight knots and sold them to the fresh-fish markets.

At the outset of the 20th century the BC Fishing & Packing Co. and Stephen Cook held the first two drag-seining licences issued for the Nimpkish River. The original owners of the resource, the Kwakwaka'wakw people, were rewarded with foreshore rents and a market for the fish they caught. But by 1927 non-Natives operated 27 drag seines and 30 purse seines in the fishing grounds in and around the Nimpkish. They were fishing every little creek and penetrating deep up the rivers. To the First Nations people money was a sort of perk. The salmon, on the other hand, were their lifeblood, and they could see them hemorrhaging away.

There were so many white men involved in what had originally been an aboriginal fishery that the Native people were forced to petition the federal government to "grant" them rights that were inherently theirs—an irony seemingly lost on those who were there at the time. When William Found, the director of the West Coast fisheries, called a meeting at Alert Bay in 1927, the First Nations fishermen came out in full force to voice their complaints and make recommendations. As a result, as Sam Hunt said, "They kicked out the whites." Only the Kwakwaka'wakw were permitted to fish on the Nimpkish, and all trollers using gillnets were banned from the river. It hardly mattered, for now seiners were catching the fish before they ever reached the river.

In 1914 the Hell's Gate slide, precipitated by railway construction in the Fraser Canyon, had decimated the Fraser River salmon runs, so the canners began to concentrate their attention on the north: the Skeena River, Rivers Inlet and Johnstone Strait. They had created a market in Europe for coho and pinks and chum salmon, as well as for sockeye and springs. Prices had gone up because of World War I; now there were several fishing companies competing in the Alert Bay area.

By the 1920s the wooden skiffs had been replaced, first by gillnet boats and then by seiners—powered boats that could fish farther and farther afield and return to the canneries with holds full of fish. Their nets didn't entrap individual salmon as a gillnet did but caught them in the encircling web of a purse seine. Seiners have been described as "the most efficient fishing craft in the world." In 1926 Sam Hunt was in Namu on a seiner that carried a load of 220,000 humpbacks. "I made $1,800 for my share," he said. "That was a fortune." And this was the return for a crewman. For the companies who were making the real money this kind of financial reward encouraged expansion. Flush with profits, the canneries built seiners at an incredible rate. Between 1922 and 1925 the number of seine boats doubled to 302 boats, and in 1926 another 104 were launched.

The Native people, once the sole providers of salmon for the cannery in Alert Bay, were now crewmen on what were almost exclusively company boats. Sometimes they made a pile of money, as Sam Hunt did, and sometimes they didn't. For their real sustenance they turned to the Nimpkish River as they had always done. The local First Nation council, headed by Jimmy Sewid, arranged for the various tribes to

Native fishermen pose with their nets at Alert Bay, circa 1918. From the collection of the author

take turns going over to the river to fish for food. It wasn't as it used to be in the days of Franz Boas, though, when there were catches of 5,000 fish; in the summer of 1956 drag seining brought in a total of 1,500. And the numbers continued to decline. "After a few years we gave that up," Sewid said. "There weren't enough fish in the river to make it worthwhile." The Nimpkish River's bounty would never return.

In 1896 the Union Steamships began to call regularly at Alert Bay. These little ships with their black hulls, white superstructures and distinctive brick-red funnels were emissaries from a larger and more civilized world. An arrival was the event of the week. Onlookers watched as winches hissed steam, and pallets of food, hardware, dry goods and supplies of all kinds emerged from the hold and were lowered onto the cannery dock. Other pallets piled with boxes of fish went into the hold and made the return journey to Vancouver. Since the Union boats were Alert Bay's only regular link with the cities to the south, they brought the mail and other messages, and an increasing number of white residents to the island. By 1912 there were 20 white individuals living in Alert Bay and 400 First Nations people. And by the mid-1920s the small red-stacked steamers that eased alongside the BC Packers dock were servicing a community that had grown almost beyond recognition.

There was an imposing residential school, its entrance flanked by two huge thunderbird totems, and a sizeable box factory, its burner belching smoke. Cook's wharf, the BC Packers wharf and several oil docks extended their rows of pilings out into the bay. There was a hospital, a post office, a telegraph station, a BC Provincial Police station, a hotel (read beer parlour) and several stores.

From the days when Emily Carr had painted Kwakwaka'wakw canoes framing a picturesque pebble beach, to the bustle of the 1920s, a transformation had taken place in Alert Bay. If the "information age" with its technical demands sometimes proves daunting to us today, consider the Native inhabitants of Alert Bay in the era just after World War I. In some 60 years they had gone from barter to capitalism, from the medicine man to the mission hospital and from a dugout canoe (albeit a very beautiful one) to a seine boat with an internal combustion engine. Like it or not they had been hurled into the 20th century.

The pioneering spirit that created that first fish-processing plant in Alert Bay was hard to recollect in the 1920s, for by then there were canneries all over the coast. In the area around Alert Bay alone—Johnstone Strait, Blackfish Sound and the mainland inlets—the consistently good runs of fish had attracted two other fishing companies. At Bones Bay in Baronet Pass and at Glendale Cove up in the reaches of Knight Inlet, pilings were driven into the sea bottom and wharves and buildings owned by the Canadian Fishing Co. and ABC Packing Co. sprang up in an otherwise untouched wilderness. And at Alert Bay the original cannery had long since been replaced by a larger one.

Wherever they were and whoever owned them, the canneries presented the same appearance. There would be a vast wharf supporting

Opposite: The Jean W *was a typical narrow-beamed 1920s seine boat. She belonged to Jack Warren, Steve's father.* From the collection of David Huson

JEAN.W

two or three two-storeyed cannery buildings and a straggle of smaller frame buildings to house the manager and his employees. There were net lofts in the second storey of the cannery buildings, and warehouses and the machinery for the canning lines below. Much like huge barns, the unfinished interiors were whitewashed, and portable Coleman lamps supplied lighting in the early days. A couple of big diesel engines, usually Fairbanks Morse, supplied power for the lacquering machine and the fish haul, and there were steam retorts to cook the cans. Quantities of fresh water were required, and they came from large storage tanks on a hill above the cannery or from a nearby waterfall. Insurance inspectors described the climatic conditions inside the cannery as "humid." "Wet" would probably be a more straightforward description.

Since Franz Boas first described the wooden building on pilings, where the fish were piled at one end and the processing was all done by hand, canneries had become mechanized. The one at Alert Bay was typical. There were holding bins for the salmon and conveyor belts to carry them from one step to the next. There was a machine called an "Iron Chink" that gutted and trimmed the fish and another that sealed the filled tins.

Some white women now worked in the cannery. From April to September, the women and the Chinese labourers sloshed around in a world of running hoses and fish slime, their feet encased in rubber boots, as they washed, packed and cooked salmon. Tasks were rigidly dictated by gender. The Native women were still preserving the catch as they had always done, but now in vast quantities that would be eaten by strangers half a world away. The beach fires with the grandmothers were fewer. Instead there were paycheques that bought corned beef and macaroni and "a case of ketchup for the kids."

The only connections with earlier times were the camaraderie of group effort and the rhythm of the season. The First Nations people were accustomed to working long and hard when the fish were running, and the fishing industry was nothing if not seasonal. There the similarity ended, however, for this wasn't a food fishery; it was a commercial enterprise and subject to market forces. In the seasons of

The Cook clan poses for a formal photo. Back row: Willy, Alice Olney, Jack and Grace Warren, Emma Kenmuir, Stephen Jr. Middle row: Herb, Gilbert, Cyril, parents Stephen and Jane Cook, Ernie, Reg. Front: Christopher, Pearl and Winnie. From the collection of David Huson

big catches and low prices, seine boats and their loads of fish were simply turned away by the companies; the fish were dumped and the fishermen left unpaid for their work. But in the years of big runs and high prices the heavy thump of the Fairbanks continued 'til eight or nine at night, and the cannery workers carried on sometimes for 12 hours a shift.

When the seiners were still disgorging salmon in the evening hours, some of the older children, including Sam Hunt's two brothers, were pressed into service. They were not ideal employees. "They got 10 cents an hour," Sam said, "but half the time they were playin' down there—teasin' the Chinamen. They'd all get fired for throwing fish guts at the Chinamen. All come home 'fired again.' A week later the cannery would be stuck for help and they'd all go back to work."

Stephen Cook was one of those who managed to take all this change in his stride. He and his wife, Jane, and their 16 children lived in a big square house on a hillside. Stephen had the wharf and the store

The Cook family fleet at Cook's wharf, Alert Bay. From the collection of Barrie McClung

he had built with profits from the timber sale, and one of his daughters was the fish buyer at a camp he established at Harbledown Island.

In 1924 Stephen had a seiner, the *Pearl C.*, built in North Vancouver, and in 1927 he ordered another, larger boat, the *Winifred C.* Two of his older sons skippered them, and the younger boys grew up on them, absorbing commercial fishing as effortlessly as the young pick up a foreign language. In time these kids became young men and bought their own boats: Chris and his brother-in-law, Steve Warren,

bought the *Kitgora*; Gilbert had the *Cape Cook*; Reg owned the *Cape Lazo*; and Herb the *CN*. This family fleet of six seine boats all flew the red and white Canfisco flag. Early on Stephen had chosen to fish for the Canadian Fishing Co. As his grandson, George, said, "Steve was born Canadian Fish and he worked for them right to the end. All the Cook family, their career was with Canadian Fish. The canneries lent money to the fishermen to keep people loyal but that wasn't really necessary with the Cooks. You were a company man and that's all there was to it."

The fish companies, encouraged by the efficiency and versatility of

The herring reduction fishery. In winter seiners exchanged their salmon nets for herring nets. Photo courtesy UBC Library Rare Books & Special Collections, Fisherman Publishing Society

seine boats, were indeed enthusiastic loan officers. More fish meant more money, and a loan kept those fishermen who wanted to own their own boats from straying to rival companies. As soon as a boat was paid off, its owner was encouraged to buy something bigger and better—thus maintaining his indebtedness. The April 1927 issue of *Harbour and Shipping Magazine* reported that "last year there were about 55 new seine boats built on Burrard Inlet and the Fraser River." In another 20 years this policy of unfettered expansion would have serious repercussions, but at the time the many new boats, each bigger and more powerful than its predecessor, were a status symbol and a source of pride.

For by this time it was obvious to all that a successful fisherman, the owner of his own boat, was also a successful businessman. A seine boat represented a large capital outlay, and as well as the boat itself there were the costs of the net and the gear. There was insurance and fuel and maintenance. To finance all this the boat's owner had to possess the elusive skill that it takes to catch fish.

He had to be enterprising, too. The flourishing fishery of the late 1920s, when Stephen Cook's first two boats were built, became the bare subsistence fishery of the 1930s. In 1928 Steve Warren, Stephen's grandson, still a schoolboy, made $400 fishing over the summer. "It was a fortune," he said. But those days had gone. The salmon were still plentiful enough but the prices for them had dropped through the floor. "In 1930," Steve said, "we got 25 cents apiece for those nice Nimpkish sockeye, and after the sockeye we went to the mainland inlets for humps—two and a half cents apiece. Coho we got 12 ½ cents each and red spring over 10 pounds we got 50 cents apiece. Anything under 10 pounds they wouldn't take and they wouldn't take white springs at all. It was hard to make any money at those prices."

By 1932 it was obvious to Stephen Cook that the family must utilize their boats better by earning some money in the winter. He conferred with his sons and they decided to try their hand at "the herrin'." They became some of the earliest commercial herring fishermen. "The Alert Bay district was great herring country," Steve Warren said. "We fished Bones Bay, head of Knight Inlet, Belleisle Sound, Sargeaunt Pass, Greenway Sound—and Retreat Pass, which was the place to catch the bigger herring. Canadian Fish liked those bigger herring. They sent them to town to make kippers."

And then, having ranged the convoluted passages that formed the mainland, they went out the north end of Vancouver Island to fish for halibut. In early February they caught herring for bait, took them into Cascade Harbour where there was a freezing plant and had them frozen into 90-kilogram blocks. By early March they were heaving around on a cold grey sea looking for halibut. "There was no quota, nothing like that," said Steve, "and there wasn't a lot of halibut in close at that time of year, but we did fairly well. We weren't highliners but we were bringing in some revenue at a slow time of year.

"We'd ice up at Jim King's and leave Alert Bay on Sunday. We'd stop at Cascade Harbour the next morning to pick up our bait, then we'd head out of Bull Harbour and fish Fisherman's Bay, the Yankee Spot, places like that. We'd head home early Saturday so we could re-ice our fish and put them in boxes and send them to town Sunday on the Union Steamships."

It was a punishing schedule. Asked what it took to be a good fisherman, Steve Warren said, "A lot of hard work."

Sam Hunt was born in 1907. "We've got what they call a live roller now," he said when he was 70, "but when I started fishing there was no such thing. When you were seining you just pulled the net over a roller, a little skinny thing. Today it sounds like a lot of hard work but in those days you didn't know any better. And then the power block came in. There was a fellow that helped us load our net with this power block. And we just couldn't take it after that. We went to Vancouver right away and had one put on—hydraulic. And then after we fished a few weeks with this thing it broke down. We could hardly pull the net in. It was just imagination, I guess. Seines that we pulled in by hand all those other years."

David Huson's grandfather, Spence, weathered the dirty thirties better than most. He had a steady job running the ice plant at the Walker Lake cannery in Johnston Channel. "My grandfather told me that some ladies had a house on the point," David said, "just where you turn to go into Ocean Falls. All the loggers used to stop there after they got paid. It didn't last long, though, because one day after the loggers had gone back to the woods the women at Ocean Falls went over and burned it down."

For those who were still trying to make a living fishing—people like Steve Warren, Sam Hunt and the Cooks—life was bleak. At the end of the season, after paying their expenses, many fishermen were broke or, worse still, in the hole. In 1936 a bitter strike for better fish prices paralyzed the coast. In Alert Bay, BC Packers tried to persuade Dong Chong, the owner of the grocery store, to deny the strikers supplies. Dong Chong, a pragmatist, knew it was the striking fishermen who were his customers and so, of course, he refused. Two years later there was another strike, again protesting the prices the canneries were paying their fishermen. Then, finally, a war came to the rescue; World War I had created better fish prices and so did World War II. Fishermen who had been earning $400 or $500 a crew share were now making $1,800 to $2,000. In 1942, a good sockeye year, Sam Hunt earned a share and a half, which amounted to $5,000. After years of depression it was almost unbelievable. "I'd never seen that much money in my life," he said.

Money like this attracted would-be crewmen, too. "In '41 the crew were getting dragged into the army," Hunt said, "and this guy Don was after me to go fishing. He worked on a piledriver before that; he's never fished in his life. So I said, 'You'll have to cook. That's all there is.' We went to Granite Bay. It was still dark in the morning. We tied the net to the beach. It was against the law [a law now changed] but we did it anyhow and set the net out. So somebody said to Don, 'Go and plunge off the bow.' No move; he's just standing there. 'Come on, go and plunge off the bow.' Finally they yelled, 'Take that plunger and plunge off the bow.' Well, Don, he grabbed it right away and ran up to the bow. He told me later he thought he was ordered to go and jump off the bow. It was still dark and cold. He said he was going to quit right there and then. He wasn't going to jump off the bow." (Up until that time "plunge," to Don, was a verb meaning to jump or dive into water; he had never heard of a pole with a cone-shaped end that "plunged" the water scaring the fish into the net.) When this confusion was cleared up, Don stayed and fished with Sam Hunt's crew for the next 18 years.

With prosperity, the gentle pace of life in Alert Bay quickened. Before Charlie Pepper built the Rainbow Theatre, he showed movies every Saturday night in Alert Bay's Anglican church hall. This was in 1941 and there was no competition from television. Because the only other forms of recreation in Alert Bay were drinking, favoured by loggers and fishermen, and bridge games and badminton, the choices of the "respectable people," Charlie's movies were the high point of the week and well attended. The films arrived in circular metal containers piled into the canvas mail sacks that Union Steamships delivered each week from Vancouver. More often than not they were shoot-em-up, cowboys-and-Indians movies, extremely popular with the audience in the days before political correctness set in. The real Indians yelled encouragement to the phony Indians on the screen. And the small boys, sitting on the wooden benches in the front row, totally caught up in the action taking place on the screen, stared up in wonderment at the moving images. When the story line got particularly exciting they shrieked and pummelled each other until the bench fell over, throwing them all on the floor.

None of the audience gave any thought to the fact that, sitting there in the dark, they were looking at Hollywood's version of the Wild West while the real Wild West was going full tilt right outside the door. There was a rickety little village outside that included, instead of the Last Chance Saloon, the Nimpkish Hotel. The steeds weren't tied to hitching posts outside the bank but were moored five deep at the string of docks that lined the shore. And since it was Saturday night in fishing season and everyone had just been paid, the good guys and the bad guys were locked in combat—rolling around on the gravel road that formed the main drag. Fuelled by alcohol and testosterone, they flailed away, barely noticed by passersby.

Few of the combatants were aboriginal. Since 1867 it had been illegal for First Nations people to possess hard liquor, and even in 1960, when they got the vote, they were still forbidden to bring it onto their reserves. The rationale behind this discrimination was the firm belief that there was a physiological difference between the Native and the white man, and that this genetic difference caused the former to react especially violently to alcohol and thus endanger himself and others.

In the early days, when a member of the First Nations community in Alert Bay got hold of a bottle of rye whisky, drank the whole thing and went down to his gillnetter and blew his head off, it was proof to the white residents that this theory was correct. The fact that anyone with a whole bottle of rye inside them might lean toward irrational behavior, if not simply keel over and die, was not considered.

And the fact that selling hard liquor to the Native population wasn't legal didn't prevent it from happening. It just meant that the market value of that liquor increased. Slipping alongside certain docks in the darkness one could find a bottle or two being passed down through a hole in the planking. Despite the law, then, it had always been possible for First Nations members to share the white man's right to fall down in the middle of the road, neglect his family and waste his money.

The size of Stephen and Jane Cook's family was something to remark upon, even in those days, but there were many other big families in Alert Bay. In the early part of the last century families were larger and closer—both physically and emotionally—than they are today. And in places like Alert Bay on the edge of the wilderness it was the Chinese and the Native families who were largest and closest. Both cultures had a long tradition of familial respect and responsibility; aside from that, however, their reasons for close family bonds couldn't have been more different.

The Chinese had left their roots and everything familiar. To succeed in this new world (and they intended to succeed, not just survive) took the combined efforts of everyone in the extended family, from toddlers to the aged. They bent to this task with cheerful but unwavering determination.

The First Nations people, on the other hand, had a connection to this land—and this sea—that stretched back thousands of years. Their roots were deep and their extended families formed a huge web of interrelated kinsmen. And their very existence had depended upon communal participation.

In a sophisticated urban setting it's possible for one enterprising individual to feed, clothe and house a family and, if required, hire a nanny to care for the children. This arrangement may be stressful enough to send you to a psychiatrist—but it is possible. But it takes more than one to kill a whale or to catch any appreciable number of salmon. It took a whole crew to raise the massive logs that frame a longhouse. An extended family was necessary to serve as substitute parents for the orphaned and the illegitimate. And only a close-knit group could fend off the loneliness of a vast country that dwarfed its human occupants. In First Nations society, ties to one's family and village guaranteed survival, and they were strong and deep.

It was not simply survival, however, that encouraged a gregarious way of life. For the First Nations people happiness was not only a large family but a whole crowd. They grew up doing their work in the company of their fellows in an atmosphere full of laughter and jokes and general good humour. The word "camaraderie" could have been invented to describe the sociability of this way of life.

CHAPTER 4

High Boats:
Miss Helen — Nick Jurincic (ABC)
W. No. 11 — M.K. Alfred (BCP)

PIONEER JOURNAL, SEPTEMBER 20, 1952

On a sunny September morning, more than 100 years after his paternal great-grandfather came to Alert Bay looking for the main chance, David Huson stood on the *May S.*'s dodger and smoked a cigarette. With no net on the drum and no fish in the hold the *May S.* had a bone in her teeth. David watched her big bow wave part the glossy water and roll away to form a lacy *V* on either side.

David's family ties stretched back to the very beginning of the community into which he was born. Wes Huson was David's great-grandfather. His great-grandmother was the handsome Native woman, Mary Ekegat. His grandfather was Wes Huson's fifth son, Spencer, and his father was Spencer Huson, Jr. It was Stephen Cook's granddaughter who would grow up to become David Huson's mother.

From the very beginning, David found himself a member of an extended family. Although the longhouses might have gone, the sense of communal living hadn't. As well as parents, there were uncles, aunts and cousins to prove to this small, rosy-cheeked child that the world was a warm and friendly place full of people who cared for you. And long before psychologists stressed the need for role models, David had them. When he was a toddler and his father went north to fish, his uncle, Jack Scow, became a substitute. He packed David down the dock to his gillnetter, tethered him to the mast with a length of line and introduced him to commercial fishing.

David Huson as a toddler straddles a bicycle (a long way from the pedals), supported by his father, Spence Jr., and his grandmother, Elizabeth. Spence joined the army in World War II, but was never sent overseas. From the collection of David Huson

From the start, life was purposeful and fascinating. And from the start it was intimately connected to the sea.

Unlike most of his contemporaries in Alert Bay, David's life had another dimension as well. His mother's brother, Steve Warren, lived in Victoria, and when David's father and his uncle were fishing in the north for months on end, David and his mother went to the city to stay with the Warrens. At three, David attended Miss Ashdown's nursery school, an establishment that reflected the mores of the day, which were perhaps more artificial in Victoria than elsewhere. Hence there was a marked contrast between the days he spent in the company of polite, tidy children who had all been taught to "share" and the months spent with the rowdy little boys who played on the beach at Alert Bay.

Alert Bay was where he belonged, however; there was no doubt of that. But days on the beach were soon replaced by days in school. By Grade 3, the regimentation of a classroom had lost whatever novelty it had first possessed for David, and he regularly terminated his school day at recess time. Until his cousin in Grade 12, under orders from David's parents, put a stop to this truancy and confined him to the school building, he spent the majority of his school days exploring a nearby swamp. When that escape route was cut off he had to wait until summer to obtain his release. It was only in the summer holidays, on the *Lake Como IV*, that he was learning what he wanted to learn—to become a fisherman.

The BCP boat Fisher Lassie *was skippered by Tommy Hunt, among others. Tommy, one of David Huson's mentors, could determine the location and density of a school of herring by the number of fish that hit a piano wire lowered into the water.* Photo courtesy UBC Library Rare Books & Special Collections, Fisherman Publishing Society

"Goin' on the boat young, that's the way it was," he said. "'Specially if you're a boy in a fishin' family." To accommodate this budding crewman, David's grandfather constructed a child-sized bed from galvanized pipe and fastened webbing across the bottom of it, and his grandmother made a mattress out of two pillows. The whole thing fitted over the storage batteries in the engine room. "Now there's all these laws about batteries," David reflects. "You have to have them all vented and covered now, but nobody knew that in those days." So he slept above the bank of batteries with the engine's metre-wide flywheel and several spinning belts right beside him. Sleep was the only thing that stilled his ceaseless activity. "My Dad was very patient," David said, "but sometimes I musta driven him crazy." In order to keep track of him the crew occasionally tossed him into the brailer where he thrashed around like the salmon who fought for their freedom in the same confines.

Aside from custom and his own preference, there was another

reason that David spent so much of his time on the *Lake Como IV.* His mother was suffering from an illness that would eventually kill her, and as a consequence, a good part of David's care was assumed by his father, his aunt and his grandparents. A family friend said, "David liked fishing, which was a good thing because he didn't have much choice." "Liked" was not really the right word, though. "Fishin' was my life," David said.

The summer he was 11 he became a bona fide crew member. Now each time the *Lake Como IV* made a set he was charged with rowing the skiffman ashore—quickly. Rowing a brutally heavy skiff that is dragging a net behind it is not exactly duck soup for an 11-year-old. David's arms were barely long enough to span the boat's two-metre beam, and the oars were heavy and cumbersome. But he had rowed for as long as he could remember and he managed—all summer long. He received a half-share of the profits with justifiable pride, and the next year, at 12, although still considered a "summer kid," he earned and was paid a full share. Now he could rightly call himself a fisherman.

The *Lake Como IV* had a power block by this time, but aside from this innovation her crew had little in the way of technology; instead they substituted experience and intuition. The time David spent fishing herring with Tommy Hunt was an example of this. Tommy was an old man then—a man who had never gone to school and couldn't read. Yet he navigated his boat all around Vancouver Island and fished with an instinctive understanding of tides and fish movements that was a gift from his gene pool. David watched him lower a piano wire into the water below the boat. Tommy could tell by the number of fish that hit the wire how many tonnes they would be likely to catch.

When technology did arrive Tommy took full advantage of it, but he added an indefinable human dimension. David remembers the crew on Tommy's boat ogling the sonar and the big balls of fish it revealed on the bottom. At 4 a.m. they were there, waiting tensely for the fish to come up. Tommy peered at the sonar. "Just go back to bed. That's it," he said. All week this performance was repeated. The other boats in the fleet began to doubt that these were fish they were looking at and one by one they left. Tommy's boat stayed right where it was.

On Thursday morning Tommy said, "They're coming up today." They caught 200 tonnes of fish.

The brothers Reg and Gilbert Cook also carried in their being an innate understanding of the water and the fish that swam in it. They knew when the fish would be coming through Blackfish Sound; they knew how far down the beach to set; they understood every nuance of the tides. The ultimate for an eager young fisherman was a job on the *Cape Lazo* or the *Cape Cook*.

At Alert Bay the year had its own rhythm, a rhythm dictated by the seasons and not so different from that of 1,000 years before. The eulachon run in Knight Inlet ushered in the spring. In May and June it was halibut season. The boats fished for salmon from June to October, and from October to January it was time to fish for herring.

The Husons lived a kilometre or two from the docks. In the summer someone would trudge up the gravel road with a fresh salmon; there would be vegetables from the garden and a big dinner for friends and relatives. In the fall it was time to hunt ducks and geese. Everyone piled onto a boat, pitched in for the cost of the fuel and headed for their favourite spots on the mainland. The deer season came next. Jack Scow always took a group on his boat to go hunting. "I can remember all the deer they'd bring in," said David. "It was like a victory thing—like the old days." The Husons spent Christmas Day at home, but on Boxing Day they took their boat and just "went cruisin' all around for a few days." It was the natural world that gave them their pleasure.

By 1950 there were 1,000 people living in Alert Bay, and the whole town was occupied with the task of building a community hall. Shares were sold to raise capital; local logging companies donated timber; and a group of volunteers, Gilbert Cook prominent among them, worked as carpenters and labourers. Within a year, white, Native and Chinese helpers had the exterior of the building completed. The maple floor was stored, ready for installation, in the cannery warehouse, and the unfinished hall made available for dances. For all its rough edges Alert Bay was still a small town in an age of innocence. There were bazaars and box socials as well as a weekly dance. Local

Jim Sewid's Twin Sisters *carried Native dancers celebrating the coronation of Queen Elizabeth II.* Photo courtesy UBC Library Rare Books & Special Collections, Fisherman Publishing Society

merchants gathered up pickaxes, shovels and gravel and repaired the road in front of their own businesses. The newspaper published "umbrella found" notices and reminders from the police that it was prohibited for more than one person to ride on a bicycle.

It was a community where salmon ruled. This was evident even on holiday celebrations. As well as a parade, a ball game and foot races, Alert Bay's May 24 celebrations always included a seine-boat race that whipped onlookers into a frenzy of excitement. Engines revved up to unhealthy rpms, the boats stormed down the length of the bay, whistles blasting and Union Jacks flying beside their fishing company flags. And in early June 1953, when a new British queen was being crowned half a world away, Alert Bay's tribute was typical. Fifty seiners and as many gillnetters assembled off Haddington Island. At 11 a.m., guided by the *Stuart Post*'s radio telephone and led by Dave Dawson's *Queen Elizabeth* and Jim Sewid's *Twin Sisters*, they streamed down the strait with flags snapping in the wind. On the

When Spence Huson Jr. retired from the fishery, Elizabeth Huson added four bedrooms to the family home, converting it to the Straitsview Rooms boarding house. From the collection of David Huson

Queen Elizabeth's stern deck an orchestra, which included the boat's skipper on accordion, played "God Save the Queen," while a group of Native dancers in full regalia danced on the deck of the *Twin Sisters*.

In 1950, 100 seiners and 1,500 gillnetters, one-third of them from Alert Bay, caught 18,272,000 kilograms of salmon in the area. The catch was worth $4,966,317. The Department of Fisheries employed three inspectors locally, plus 11 patrolmen, one department boat, 11 chartered boats and a chartered aircraft. And the Alert Bay newspaper devoted a good part of its space to changes in fishing boundaries and weekly catches, and expressed much concern over the state of the spawning streams. In August 1952 it reported a black mass of sockeye packed into the pools below Karmutzen Falls in the Nimpkish River. The paper also reported the results of the annual seal hunt. In 1951 two fisheries patrol boats from Alert Bay shot 120 seals in Knight Inlet and Mackenzie Sound in order to protect the salmon runs. Long before fish farms the seals' appetite for salmon was their undoing.

That same year, 1950, David's grandfather celebrated his 74th birthday. There were 18 for dinner and Spencer reminisced about going to Fort Rupert with his father in a big canoe paddled by eight Native men. In the age of seine boats, it was not something any of the other family members could relate to. Now virtually blind, David's grandfather spent his days in his little skiff powered by an equally small Seagull outboard. As long as the family would let him, he putted across to the Nimpkish River to catch crabs or fished the tides that swept past Alert Bay. Occasionally, when eyesight or engine trouble delayed his return, David's father set off in the *Lake Como IV* to rescue him.

David's grandmother, meanwhile, had four more bedrooms added to the family house and took in roomers. She boarded young teachers and nurses, and their employers could rest easily knowing that Mrs. Spencer Huson kept an eagle eye on her charges' after-dark activities. David and her other teenaged grandson, hormones surging through their bodies, were greatly attracted to these young women, but they could forget any fantasies they entertained. "There was no hanky panky in her house," David said. "You couldn't even go to the washroom without finding her standing there by her bedroom door."

It was a strenuous routine for a woman who was no longer young. As David said, though, she made it all work. Her grandchildren were enlisted as helpers, the girls to wash dishes and the boys to fill the oil tanks and carry the grocery orders. Then to augment the income from this labour-intensive but not particularly lucrative business, Mrs. Huson took the family boat out every weekend on a "shopping run," carrying loggers' wives from Beaver Cove and Port McNeill to the shopping centre of Alert Bay. David was her crew. There are no stereotyped roles in places like Alert Bay, and none of their passengers found it surprising that this ferry service was operated by an elderly woman and a boy.

When her husband's eyesight first failed, Mrs. Huson had switched, without skipping a beat, from housewife and mother to mariner. Having raised eight children she now became captain of the *Shumahalt*. Using her eyes, she and her husband ranged the coast packing fish from as far north as Rivers Inlet. For a woman who could negotiate the tides in Seymour Narrows with a load of fish, the

Elizabeth and Spence Huson aboard the Shumahalt *at Alert Bay in 1958. When Spence's sight began to fail, his wife took over as skipper.* From the collection of David Huson

shopping run was hardly a challenge. Early each Saturday morning she and her grandson climbed aboard the boat. David cranked the engine's heavy flywheel, and in good weather and bad the two began their circuit of Johnstone Strait. They went to Beaver Cove first, picked up their passengers and a fistful of grocery orders and delivered them to Alert Bay. Then they picked up a second group of shoppers from Port McNeill. By the time these had been off-loaded in Alert Bay the first group was ready to return to Beaver Cove. Then the boat took the Port McNeill shoppers back and finally headed home to Alert Bay. That evening the *Shumahalt*'s captain cooked dinner while her engineer argued with his father over the choice of a TV channel. David was partial to wrestling.

Mrs. Huson's passengers were the employees of huge companies like Alaska Pine, Crown Zellerbach and Canadian Forest Products, which had bought out the small independent logging companies and had literally changed the face of the earth. Now instead of an A-frame, an ancient "donkey" and a few board-and-batten shacks on floats, there were big camps slapped down in the middle of nowhere: prefab bunkhouses, cookhouses, married quarters, machine shops and offices laid out in rows. There were steel spar trees now, and a line of gigantic logging trucks with tires as tall as a man. These companies were the possessors of timber licences for vast tracts of land, and they proceeded to lay bare mountainsides and denude whole valleys. Now when it rained, a torrent of mud and tree stumps poured down the Nimpkish River. Their payrolls brought more money to Alert Bay businesses than they had ever seen before—and their logging practices sowed the seeds of destruction.

The fact that Alert Bay was booming didn't make it a better place to live. The responsible citizens of the community did their best to counteract the examples of financial and alcoholic excess that surrounded their children by building a bulwark of middle-class institutions. They organized a PTA, established Guides and Boy Scouts, and held pageants and teas. They canvassed for the Red Cross, formed AA groups and attended church. But theirs was a rough and thorny path. In 1958 the mimeographed pages of the small local newspaper, the *Pioneer Journal*, published a "page of protest" that summed up the situation. The editor wrote:

> We understand that unless evidence is given of wide-spread opposition to the proposed liquor store here the LCB will proceed with their plans. Any doubts we may have had about opposing this were entirely removed during the Easter weekend.
>
> The street fights and obscene language during these three days were disgusting and desecrated that sacred celebration. How can anyone in their right mind think of adding to this by providing an additional source of liquor. It was excessive drinking that closed the Parish Hall for dances and that threatened to close the dances at the Community Hall.

This anguished protest failed to stop the wheels of progress and the liquor store became a reality. The only thing that changed over the years was that drinking opportunities became more democratic. By 1951 the aboriginal population of Alert Bay, hitherto denied access to beer parlours, had won the right to frequent these places and get bombed out of their minds.

The results were sometimes tragic. The January 1958 edition of the *Pioneer Journal* noted that a couple from Smith Inlet had received a six-year suspended sentence for the death of their three-year-old son. He fell off his parents' gillnet boat on the night of October 20 and drowned. His parents were in a police cell at the time, "having been apprehended in a drunken condition."

In 1956 David Huson left this tough little town to go to Victoria to high school. He lived with his uncle and aunt there, and this family atmosphere made the move to the city easier than it might have been. His easygoing nature helped cushion the culture shock, as well. Still, it was an upper-class neighbourhood; the children of doctors and lawyers made up a large part of the student body. Their sophistication, social skills and prowess on the football field made them the school's inner circle. Those outside this charmed circle, including David, gathered in the bushes at the far end of the football field and smoked cigarettes.

The curriculum, too, left much to be desired in David's eyes. Logarithms and the poetry of John Donne could not, despite his best efforts, be transformed into anything very useful. Sometimes he didn't bother to get up in the morning. On the days when he played truant, his aunt opened his bedroom door and yelled threats, but they both knew that her heart wasn't in it. Later that day they would sit together at the kitchen table laughing and drinking coffee.

By now 14, David could handle a 20-metre seine boat. He knew where the tide poured through passes and channels. He could mend nets. He had his own money, earned with his own hard work. At Alert Bay he and a cousin copper-painted the *Lake Como IV*'s hull, and as payment his father chartered a seaplane that dropped the boys off on a remote little lake. There was a shack on a tiny island there and a water-filled rowboat. The boys caught trout, cooked them over an open fire, tried to paddle to shore before the rowboat sank—and had a thoroughly enjoyable time. It's doubtful that his classmates in Victoria would have enjoyed such an experience, let alone survived it without incident, and their parents would have felt it irresponsible to leave two boys alone in the woods. These, of course, were the same parents who gave cars to their own teenage sons and set them loose on the highway with barely a second thought. The two worlds had nothing at all in common, and there seemed nothing in one life that was transferable to the other. In Alert Bay David was already a fisherman; in the Victoria high school he was an outsider.

The one bright spot in this uncongenial educational experience was David's friendship with Barrie. They met out there at the end of the football field and soon became inseparable. They spent their after-school hours consuming as many chips and Cokes as their finances allowed and then invariably ended up sprawled on the rec-room floor at David's adoptive home watching *American Bandstand* on TV.

Barrie had a vast paper route, and each morning at 4 a.m. he tapped on the window of David's basement bedroom. David, used to being roused at dawn to catch fish, rolled out of bed and the two of them wheeled through the silent streets of Victoria throwing rolled newspapers onto porches.

CHAPTER 5

High Boats:

Chief Tapeet — Walter Hunger (BCP)
Anna M. — Alfred Hunt (ABC)

Pioneer Journal, August 19, 1964

There is a photograph of Barrie McClung and his parents taken in 1946. His mother, a slim brunette, holds her son in her arms and smiles at the camera. Beside her stands Barrie's father, a big burly man wearing rimless glasses and a navy-blue suit. Barrie is about two—and already he looks anxious.

Barrie was adopted in infancy. The Victoria social worker who placed him must have congratulated herself, for she had found a home with what she described as a "literary" family. It was Barrie's paternal grandmother, Nellie McClung, who provided the basis for this adjective. She was a remarkable woman. She wrote 16 immensely popular books and many magazine articles, but her writing was only part of her achievements. A passionate advocate for social justice, she was instrumental in obtaining votes for women and ended up becoming an MLA. But she was not, as her formidable accomplishments might suggest, a joyless do-gooder. She pursued her agenda with the humour and pragmatism that had helped her bring up five children. Her feet were always planted firmly on the ground—literally and figuratively. Brought up on a farm in Manitoba, she had struggled mightily to get her own education, and encouraged her own children to take advantage of their more readily available educational opportunities. But they should not do this to the exclusion of having fun, she stressed.

And they did have fun. It took World War II to break the spell. Barrie's father joined the Seaforth Highlanders, although he didn't get to Europe because of a leg injury. His mother attributed the cracks that later appeared in the family's edifice to this wartime experience. In truth her son probably had a drinking problem even before he enlisted. He certainly had one when he was demobilized, and this was particularly distressing for his mother. Although her own father was a model of sobriety, Nellie had seen the privations caused by drunken husbands and fathers. As a consequence, she was a teetotaller and a forceful orator on the evils of alcohol. Her son took care to hide his problem from her, but she was not a fool and he lived with her unspoken disappointment.

Barrie's father had married in the mid-1930s but he and his wife remained childless. It was whispered discreetly that his wife was unable to have children. The blame for this deficiency seems to have been entirely on her shoulders, and one wonders if that was the case. At any rate they adopted Barrie and later a daughter, and while his grandmother remained alive things remained on a more or less even keel. "She was the glue that held it all together," said her grandson.

As a grandmother she was a great success—kind and warm and full of fun. She had a loving marriage and Barrie remembers a big, unpretentious house surrounded by fields of vegetables and flowers, church on Sundays and family dinners afterward. At her death, however, the cracks in the family's facade widened into yawning chasms. With his mother gone, Barrie's father no longer felt obliged to present even a token appearance of sober responsibility. Instead he gave his full attention to alcohol. He held a series of menial jobs from which he was inevitably fired, and his family lived on the edge of a financial abyss. As is so often the case, the worse his behaviour became, the more his long-suffering wife appeared a paragon of virtue. Caught in a spiral of self-hatred he vented his frustrations on his family.

Barrie's mother had not only married into a family of some prominence, she herself had come from a well-connected clan. For this reason, perhaps, she was loathe to abandon her marriage. In the

1950s divorce was still uncommon and she feared "disgrace." As well, she feared a loss of social position and entry into circles that would not welcome a single mother. So she did not confide in anyone. Even her best friend only knew that there was "a bit of a drinking problem." She chose instead to keep up appearances, which was not all that difficult in an age when "nice people" only whispered the words "pregnant," "cancer" or "alcoholic." As recently as 50 years ago a woman's husband and children provided her status. As a young woman from a "good" family who had married into another "good" family, Barrie's mother enjoyed this status, if little else, and she chose to hang onto it.

Barrie was dressed in knee socks, short grey flannel pants, a navy blazer and a little cap and sent off to private school. After school he sat at his music teacher's grand piano and struggled to get his small fingers onto clusters of keys. On weekends he sat in the darkened Royal Theatre and watched an opera unfold or accompanied his mother to garden parties where they mingled with ladies in hats and "afternoon dresses." His education, he surmised, was paid for by his grandparents, for after they died he was transferred to a public high school and the music lessons stopped.

Money, or rather the lack of it, had by now became a major concern. His mother, seeking some (unobtrusive) way to earn a few dollars in "pin money," smocked endless baby dresses to sell. Before long, however, she found a venture more in keeping with her exceptional organizational abilities. She was a founding member of the Queen Alexandra Solarium's Junior League, which became the Cerebral Palsy Association of Lower Vancouver Island and exists today as the G.R. Pearkes Centre, a large professional facility for handicapped children. For years she was the administrator of the centre, drawing a very meager salary considering the effort she put into the job. (She was outraged when she retired after 20 years and her replacement was paid $80,000 a year. After all, she said, this was a charity.) She could have earned far more as an executive secretary but again the mores of the times prevailed. This was the heyday of the Junior League; good works were an acceptable occupation in her social circle—a career was not. So good works were the route she chose to take.

As her responsibilities increased so did the time she spent away from home. Her husband, cast unwillingly in the role of childminder, expressed his rage by bullying his son. And the resentment was returned, for Barrie despised this man who hid bottles in the basement and ran up bills for Aqua Velva, and he made no effort to hide this. He wanted, above all else, a father who had a job, who had the money to buy Barrie the Little League baseball uniform he coveted and who would come to the games Barrie longed to play. Their relationship was not improved when Barrie returned home from school one day and discovered that his father had sold Barrie's piano, a gift from his grandfather, to the milkman. Father and son, then, existed in a haze of mutual antipathy.

To escape this atmosphere, Barrie, like his mother, spent as much time as possible away from home. By the time he was seven he was building little driftwood shacks on the nearby beach where he could sit and read in peace—evicted only when rain penetrated his leaky roofs. At home he escaped his father's anger by barricading himself in his room. His security system consisted of a bar across the door and all the dresser drawers stretched end to end to the opposite wall to hold it in place. Once when his father threatened to break down the door, Barrie climbed up onto the dresser and let himself out the window. He turned up, shaken but polite, at the neighbour's back door.

"Can Bobby come out and play?"

"No, of course not, Barrie. We're having our dinner." Irritated by this interruption, Bobby's mother didn't think to inquire why Barrie himself wasn't at home at his dinner table.

It wasn't until Barrie was 13 that he met David and discovered that this new friend lived only four blocks away. David was calm and good-humoured, two traits that Barrie found in short supply at home. And when he was invited over to the house of David's uncle he found a whole family who were calm, good-humoured—and endlessly hospitable. Barrie had already determined that the less time he spent at home the happier he would be, so he spent all his after-school hours in David's company. They ate chips and drank Coke, listened to records and smoked when they could afford it. Occasionally, when it seemed prudent to do so, Barrie stayed overnight, sleeping on the floor beside David's bed. His hosts accepted this arrangement with their usual good nature, and Barrie was too young—and too stressed—to consider that this might be an imposition.

As the Christmas holidays approached, Barrie learned that David would be boarding one of the Union steamships to sail home to Alert Bay for a warm and happy reunion with his family. In contrast, the Christmas season at Barrie's house was a time fraught with perils. The gallon of wine, bought for guests, was consumed by the host in a couple of days, with resulting carnage. So Barrie, who was now 14, made a decision. He went through his little stash of crumpled bills and counted out $50 that he had saved from his paper route. This was his stake. He proposed to his mother that he, too, should go along when David left for Alert Bay. Having been adopted into one family, he was now choosing another and shamelessly attaching himself to it.

On their next visit to Victoria, Barrie's mother invited Spence and Mary Huson over for afternoon tea in order to check out their character. Since they appeared to be the fine people they were, she gave her permission.

David's father presented only one condition. When asked by his son if Barrie could come to Alert Bay his father said, "Sure, sure, but first there's something you gotta do, David. Your mother wants one of them little Scottie dogs. Look around and find a place where you can buy one and bring it up with you."

David's uncle drove them down to the stately CPR dock in Victoria to catch the midnight boat to Vancouver. The two boys and a not-yet-

Opposite: The Union Steamship Catala *served Alert Bay, bringing mail, groceries—and David Huson's wide-eyed Victoria schoolmate, Barrie McClung.* C. Kopas photo

CATALA

toilet-trained puppy spent the next day with Barrie's aunt in Vancouver. That night (with some relief, one might imagine) she drove them down to the less-impressive Union Steamships dock where they boarded the SS *Catala* for the 9 p.m. sailing. For the puppy, who embarked on these voyages hidden in David's jacket, and for Barrie, it was the beginning of a whole new life. Neither of them ever returned to their homes in Victoria.

As the *Catala* neared Alert Bay, David and Barrie, fidgety with excitement, joined the little knot of people pressing against the rails of the main deck. They waited as the gangway was hoisted into place and secured by the deckhands, and finally they burst through this exit and tumbled onto the government dock. Going down to the dock to "meet the boat" was one of the rituals of the coastal communities. Along with the whiff of smoke from the Union Steamships' brick-red funnels was a whiff of city life: mail sacks, freight, passengers in their unfamiliar city clothes. All of David's family were there—mother and father, sister, aunts, uncles, cousins—a warm, jovial group who accepted Barrie into their midst as if his presence was the most natural thing in the world. And what a world. For Barrie, who had spent many of his after-school hours barricaded in his room, it was a revelation. The swaying masts of fish boats were everywhere. This was a world full of people and boats in almost equal number, it seemed. In time he would be able to identify each boat when she was only a silhouette on the horizon and would know who skippered her.

Alert Bay had existed for almost 100 years at that time. Before the turn of the last century the shore was lined with traditional First Nation longhouses, each entrance decorated with a family crest. And all along the beach canoes were pulled up past the high-tide mark. By the time Barrie arrived there were many conventional frame houses on the narrow, cleared strip of land that surrounded the bay. And as the salmon money poured in, new businesses lined the water side of what was now a gravel road: grocery stores, restaurants, a couple of beer parlours, a shipyard, drygoods stores and, finally, a bank.

The first white men who arrived in Alert Bay in the late 1800s were largely of British extraction and brought their rigid class system with them. Despite the fact that many of them were missionaries, they did not subscribe to the notion that all men were created equal. The First Nations people, as their name implies, were the ones who first occupied the coast. That didn't deter the Anglo-Saxons, however, from assuming an attitude of superiority. If, at the close of the century, the ideal was to be white, then the ultimate was to be British; from the walls of Alert Bay's few living rooms, portraits of the British monarchy stared stiffly out at a world they would have found incomprehensible. But by the time David and Barrie arrived in Alert Bay, people from a dozen different countries had settled there; it was no longer possible, in that practical life, to be a bigot. Instead the population was divided into "decent guys" and people who were "no good." It was a useful division created of necessity by the owners of many small businesses. Very simply, people who were "no good" were irresponsible and didn't pay their bills; "decent guys" did. And it was a small town—too small to easily escape your reputation.

This was the Alert Bay that Barrie was seeing for the first time. Within the first few days he was surprised to find that this village was different in yet another way from the city he had left behind. In Victoria the focus was on getting a good education, finding a good job and making good investments. All these things provided you with security. But in Alert Bay, for the Native people at least, the resources of nature and the support of family provided security. As long as the seasons—and your relatives—endured, there was a need to work but not to search for a job. There was no need to get up early if you didn't feel like it or to agonize over an illegitimate child. Instead the focus seemed to be on having a good time—or as good a time as life permitted. And having a good time required that your activities be carried on in the company of as many friends and relatives as possible. The Christmas tree expedition was a good example.

Barrie had always assumed that Christmas trees were purchased from some cheerless corner lot. But not at Alert Bay, it appeared. David's father had always cut Christmas trees for half the village. Now, a day or two after the boys' arrival, friends and relatives piled aboard Spencer Huson's boat and set off to get a load of trees. Suitable candidates were searched out by following the shoreline so closely that it seemed any minute they would be on the rocks. Then, having reached a consensus regarding the best specimens, some of the participants clambered ashore, chopped the trees down and loaded them onto the back deck.

There was more excitement to come. While the expedition was busy cutting trees the wind had increased. By the time they entered Blackfish Sound on their way home they found themselves in a raging southeaster. Some uncle or cousin fought his way out onto the back deck and lashed down the Christmas trees, and the kids on board were sent down to the fo'c'sle. "That's the law," said David's father. "Go." At this point Barrie discovered that he possessed an invaluable gift that contributed greatly to his enjoyment of this new life. He didn't get seasick.

Lying on one of the top bunks in the fo'c'sle, breathing the thick smell of hot diesel oil, nearly deafened by the pounding of the engine, Barrie experienced the rise of the bow and then its sudden thud into the trough. Each time the *Lake Como IV* did this, the whole length of his body was raised from the mattress below and then dropped again. It was a neat sensation. The *Lake Como IV*'s small portholes with their thick glass provided just enough dim light to illuminate the boat's interior. Periodically this illumination was made even dimmer by the chaotic tumble of water outside. He and David, heads together, peered through glass streaming with water. "When you see nothing but green out there," David yelled over the noise of the engine, "that means we're not going over a wave, we're going through it." Then noting Barrie's awe, his companion's mischief surfaced. "Watch for fish," he said.

While David was at school in Victoria, his mother in Alert Bay had charge of his fishing earnings, and she doled them out to him $20 at a time. When this allowance arrived he and Barrie—the latter financed by his long paper route—abandoned all thought of an education. "We were too rich to go to school," David said. Instead they ate quantities of fish and chips from the Olde English Fish and Chip

Shop and went to the movies. On a visit to Victoria that fall, David's mother discovered that her son seldom graced Oak Bay Junior High School with his presence. She demanded better attendance in the next term and David promised her that things would improve. "But I had no intention of going back to school," he said. "Fishing was my business." In a few months the fishing season would begin and David was ready—anxious, even—to take his place among the full-time fishermen. So when the Christmas holidays came to their inevitable end, it was clear that David wouldn't be going back to Victoria. His school days were over.

This development had a marked effect on Barrie's plans. He had originally intended to stay in Alert Bay for the two weeks of the holidays, but now he too was vigorously opposed to the idea of returning to Victoria. Weeks passed and he stayed on, writing letters to his mother to ask for money. He was about to turn 15, and he felt that if he could hold on until the salmon season opened he could also become a fisherman. And from that moment on he would be independent.

It rains in Alert Bay a great deal more than it does in Victoria. Since the end of October, veils of rain had swept over the village, and it continued to pour down with varying degrees of intensity for the next six months. No amount of moisture could dampen Barrie's enthusiasm, however. He and David did chores for the latter's hard-working grandmother and gathered in the harsh fluorescent light of the Chinese cafe for chips and Cokes. They played crib and went to the movies.

The village's fishermen, meanwhile, were winding up the herring season and preparing to fish for salmon. They mended nets and repaired engines, and in this pause between fishing seasons they watched TV and caught up on their sleep.

In one of these fishermen, David's father, Spencer Huson, Barrie had found a surrogate father. "He was larger than life to me," said Barrie. "I never heard him say a critical word of anybody. He was so patient with kids, always trying to teach you and encourage you."

The rough and ready Nimpkish Hotel, where a mischievous David Huson pushed his city-slicker pal Barrie through the taproom doors and yelled, 'This guy here says he can take on anybody in the house.' Photo courtesy BC Archives, E-05281

Laughter and jokes were an ever-present part of this new life. And minor mishaps remained minor. Life was not to be taken too seriously, Barrie learned, but to be enjoyed. He went up the river at Kingcome with David and his father in a little boat powered by a four-horsepower Seagull motor. It was a primitive little outboard with a cord to start it. On the way down the river something happened to its internal organs and it started to heat up. David's father was not perturbed. He filled his bailing can with water and dumped it over the engine. All the way down the river he continued to pour water on it, laughing at the foibles of the internal combustion engine.

Any incident was excuse for an outing. For example, the news that a fisheries patrol plane had dropped from the sky into the sea—fortunately with no loss of life—prompted a trip down the strait to the site of the accident. Down by the Pig Farm, a little bight of land on West Cracroft Island so named because some overly optimistic homesteader had attempted to farm there, this aerial police car had roared in low over the hilltop, caught a downdraft and plunged into the sea. Its bright yellow fuselage and one crumpled wing were clearly visible beneath the surface of the water. David and Barrie viewed this unusual spectacle with delight, and David reflected that it served the fisheries right for spying on unwary fishermen who were unfortunate enough to have their net tied to the shore (then illegal).

For David, mischievous by nature, the greatest fun of all was introducing Barrie, a city boy from Victoria, to the vagaries of life on this seedy frontier. David's upbringing had been warm and loving—but not sheltered. He was familiar, for example, with beer parlours and their clientele, and Barrie was not. David chose 11 p.m., when the action in the Nimpkish Hotel was at its height, and then suggested that they have a peek through the swinging doors at its entrance. He flung them open and shoved Barrie into a large beaten-up room thick with cigarette smoke and garrulous drunks. "This guy here says he can take on anybody in the house," David yelled. He shut the doors and leaned against them from the outside so that the flustered teenager inside was momentarily trapped.

Outside the entrance to the Nimpkish, men leaned on the railing of the dock and smoked the last centimetre of their cigarettes. They had grimy hair and missing teeth, and their layers of frayed clothing did nothing to protect them from the rain. Barrie discovered that they were the customers of a thriving black market, men who regularly drank up all their wages and left their families destitute and were consequently banned from Alert Bay's two beer parlours. Their names were on an interdiction list circulated in the local beer parlours and the bartenders refused them service. So they stood outside in the wet dark waiting to buy liquor from others. If Barrie and David could get their hands on a couple of bottles of beer or, better yet, a bottle of rye, they could sell them to these customers for double the retail price. A transaction like this financed any number of Cokes and opened Barrie's eyes to the opportunities that existed in this new milieu. It was, he reflected, a complete change from opera and garden parties.

The Kitgora *during a sea trial prior to her purchase by Canfisco in 1918.* From the collection of Steve Warren Jr.

With the herring season over and preparations for the salmon season under way, crews were being assembled. David's dad had his crew, and for the first time David, with four years of experience under his belt, would be a full-time crew member, not just a summer kid. The other skippers had their crews as well, and even Barrie began to realize that an inexperienced 15-year-old was not in great demand. It was at this point that Steve Warren's *Kitgora* arrived. Steve had come from Victoria for the Johnstone Strait opening, bringing most of his crew with him. The one or two spaces that remained were promised to locals. His crew, too, was complete. But there was Barrie. Steve Warren was too kind a man to ignore the hopeful teenager circling the outer edges of all this purposeful activity. He not only made a space for Barrie but gave him half a crew share.

"I didn't know anything," Barrie said. "I couldn't be a skiffman. I didn't know anything on the deck and still Steve took me on." His eagerness, however, outweighed his lack of experience.

"Barrie loved his new job," his skipper said later. "He's kind of a nature person, and if you're that way you get to be a good fisherman pretty quick."

The crew must have shared their skipper's opinion for they had a few private conversations in their native language and informed Barrie that they had decided he was to have a full crew share. This was money out of their own pockets, but there was not one word of complaint from any of them. Barrie never forgot their generosity and good nature or the debt he owed Steve Warren.

Barrie was neither the first nor the last of Steve's protégés. Steve had given a succession of youngsters their first jobs in the fishing industry. Many of his crew members grew up to become skippers themselves. "Steve," Barrie said, much later, "you would have made a lot more money if you hadn't taken on all us young kids."

"That's quite a fact," said Steve in his slow, deliberate way. "Sometimes those little buggers ... I remember Oliver—David Oliver. I remember he was sittin' on the side of the boat and all of a sudden he disappeared. There was quite a lot of excitement before we got him back on board." There was a long pause and Steve smiled. "Now he's general manager in one of the big hotels in Montreal."

"And then," Barrie said, "do you remember the weekend one of us crawled through the passthrough from the galley to the wheelhouse and turned the radio over to broadcast band and listened to music from Vancouver all weekend?"

"I came down on Sunday—opening day—and the engine wouldn't start. Batteries flat."

"We were in deep shit."

"You were that. I had to get a tow to turn the propeller. Pulled her along as fast as we could and then gradually Joe Planes closed the compression cocks and we finally got her started."

For Barrie this first season was a never-ending adventure. He hauled his skinny adolescent body out of his bunk before dawn and fell back into it at 1 a.m., still wearing most of his blood-smeared clothes. He stood at the base of a bluff beside a tie-up tree, listened to the hollow call of a raven in the deep woods above and was

No matter how big the fish, no matter how much excitement this new life showed him, Barrie maintained his air of gravity. From the collection of Barrie McClung

overwhelmed by the beauty of his surroundings. He learned that there were any number of ways you could get yourself drowned. He watched a net tear itself to shreds and then watched men spread it out on a dock and mend it—laughing and joking as if they were not losing money by the minute. As the seiner rounded a point he saw a cannery appear out of nowhere: a wharf, a vast shed and hectic activity. He peered inside the big building and saw clanking machinery and rows of girls, their hair held back in bandanas, packing fish into cans. At the end of it all he had become a fisherman.

By then the first of the post-war technologies had appeared. Radar meant an extra set before dawn. Everybody was really pulling in the fish. The next year Barrie's crew made some calculations and decided that when the net was in the water they were making $90 an hour. A 16-year-old kid, Barrie told himself, making 90 bucks an hour. He was having so much fun he didn't know what to do with himself.

Although excitement boiled up within him, Barrie took care to cover it with gravity. Perhaps earlier disappointments had left him with the feeling that good fortune must be hidden lest it be snatched away. A picture of Barrie holding up a gigantic salmon shows the teenager staring straight into the camera with an expression of deadly seriousness.

Each Saturday, back in Alert Bay, the fishing company paid the crew three-quarters of the money they had earned that week. By Sunday most of that money was gone. The small stuff went to the kids. When the crews came out of the cafe they threw all their change to the kids in the street. It wasn't a patronizing gesture; it was a tradition of the potlatch. The message was clear: "In a couple of years you, too, will be fishing and making money like this. Remember to be generous with it."

There was no end to it, Barrie reflected, no end to this flow of money. Every year they would make more sets 'til they were all millionaires.

CHAPTER 6

High Boats:
Dominion No. 1 — Mel Stauffer (BCP)
Frank A M — Henry Seawid (ABC)

PIONEER JOURNAL, OCTOBER 10, 1951

"Here, see that boat coming?" David asked. "Well, stay on the outside of him and go straight in line with that point. Just stay off the kelp bed."

The *May S.* approached a sheltered bay and a patch of green paler than the surrounding trees. Here the forest was cleared away hundreds of years ago and in its stead there's a tangle of grass, moss, salmonberry bushes and blackberry vines. And this vegetation meets not grey granite but a curve of white beach—a dazzling beach that, from a distance, looks almost tropical. This clearing, with its beach of broken clamshells worn smooth by years of waves, is Mamalilaculla, one of the ancient First Nation villages hidden among these islands. Seventy-five years ago the clearing here was larger; the vegetation was held at bay and the path in front of the buildings worn bare. And 75 years ago the abandoned longhouses still stood. In the dimly lit interiors you could find bright blue trade beads in the corners of the earthen floors. Now only a few massive logs mark their locations. Beside them tall frame houses with mossy roofs tilt into the undergrowth, their vacant windows staring out to sea. Soon the rain forest will engulf the row of buildings that surrounds the small bay, and only the snow-white beach will be left to attest to the fact that people lived here for thousands of years.

Interior of a Kwakwaka'wakw longhouse. From the collection of the author

The *May S.* dropped her anchor into the glassy water and the engine's intrusive noise was stilled. In the silent little bay the *May*'s crew sat in the sunshine with their thoughts.

It had taken perhaps 1,000 years for the sea to cleanse and crush this pile of odorous refuse and turn it into pristine beach. For all those years a primitive lifestyle had endured here fed by that same sea. But in the early 20th century, when the frame houses were built, the seeds of capitalism had been sown. Change was everywhere. There were metal fish hooks and lead sinkers for sale in the cannery store in Alert Bay as well as condensed milk and spongy white bread and chocolate bars. There were those little open boats drag seining in the Nimpkish River, and then gas engines that spared their possessors the labour of rowing. As fishing boats got larger, acquiring cabins and bigger engines, their territories expanded. They fished down Johnstone Strait then and in Blackfish Sound—anywhere a gas engine could take a boat in a couple of hours. By 1927, when Kishi Bros. boatyard built the *May S.*, both the fishing industry and the fishing fleet had exploded. Big wooden seiners, with dependable diesel engines and

Hauling in the seine net by hand in Blackfish Sound, circa 1941. From the collection of the author

a turntable on the stern piled with net, dotted the waters of the BC coast.

The fishermen who manned them came from those countries with a fishing heritage: Yugoslavia, Norway, Japan and Italy. And, of course, they came from right here where it all began. The First Nations people had fishing in their blood and in their bones, for it had meant their survival. Fishing had also provided them with an easy camaraderie and the excitement of the hunt. Now, in the years after 1900, they carried all this with them into a new era. Like most coastal fishermen, they were infected with "salmon fever," and it was an ineradicable disease. Young bucks, roaring around in seine boats with the engines cranked up to the red line and the radios blaring cowboy music, were having the time of their lives. They grew into adults who were familiar with every rock and passage on BC's complicated coastline and with its strong and eccentric tides. When the fish were running they worked—cheerfully—to exhaustion. And if, when the liquor laws changed, they drank away their money on Saturday night, they knew that by Sunday evening at six they would have to be painfully sober and ready to earn some more.

For it all began on Sunday morning. On silent Sunday mornings in salmon season the fishing fleet groaned, shook itself awake and prepared to face the first day of the week. The "opening" was at six that evening. By 10 or 11 a.m. the grocery stores in Alert Bay were filled with jostling fishermen ordering meat, milk, loaves and loaves of "4X" white bread, potatoes, onions, canned fruit, jam, coffee, tea, eggs, bacon and a dozen other things, many of which were of dubious nutritional value. In the days before and during World War II, Dong Chong, the brisk and pleasant owner of the largest grocery store, filled orders and packed these mountains of food into cartons. He was assisted by his astonishingly capable children, several of whom were not a lot taller than the store's high wooden counters. The name of the boat was scrawled on the sides of the cartons with a waxed pencil, and they were loaded onto a decrepit old truck that stood outside the warehouse door. The driver who navigated this snorting old Ford down the gravel road to the docks was universally known as "Uncle." The Chinese in Alert Bay were industrious and businesslike, and their families operated as a unit. The Dong family was typical; from the youngest members to "Uncle," all were an integral part of the business.

By the 1950s Dong Chong's store had moved to a new building across the street and become a supermarket. There were no longer wooden counters, ancient scales and an adding machine with a handle that one yanked down after each entry. And Dong Chong's five children were no longer serving customers; they were now at university studying medicine, pharmacy, engineering and commerce. But on Sunday mornings the fishermen still shopped there, and cartons of groceries were still delivered to the dock and handed aboard each boat.

On the boats, engines were by now thumping away—as were the heads of the crewmen—and lineups for fuel and ice were forming. By

afternoon the boats were streaming out into Johnstone Strait heading for their preferred fishing spots: Splash Island, Izumi Rock, the Bluff, the Merry-Go-Round—nooks and crannies, rocks and bluffs, beaches and bays not identified on the chart but known by the fishing fleet that named them. They would fish off Pulteney Point, Soldier Point, No-Use Point, Hyde Creek, Dago Bay, the Kelp Patch and Green Island. The ones who favoured the strait below Beaver Cove fell in down a three-kilometre stretch. Farther south, past Robson Bight, another collection of boats dotted the shoreline.

Imagine yourself, a kid like Barrie McClung, taking part in this, your first set. Unlike David Huson, you have no fishing heritage. You didn't "go cruisin' around" as an infant or spend your toddler years tethered to a gillnetter's mast. That you have a lot to learn about this exhausting, intuitive and dangerous job is evident. That you'll have to learn it at speed is also clear, for skippers are seldom patient and instructive in the heat of the moment. First, however, your boat must secure its spot.

No matter which area your skipper favours, the protocol that governs the queuing up for sets is strict. The second boat begins the process by ensuring its position. "You first here?" the skipper inquires of the boat ahead of him. Assured of his place he can give the next turn to number three. And so it goes, the last in the queue the only one free to give the turn. When the 6 p.m. opening arrives, the first boat sets and tows. By convention he's given 20 minutes and then he starts to close. Technically, when he closes up the next man has fishable water, so if he's a nice guy and everybody's friendly, number two can set. Once in a while somebody tends to jump the gun. "You're too close," the radio barks, and rows break out. But most of the time this gentleman's agreement works well. Everyone knows everyone else and many are related; if someone breaks the rules this black duck can be punished by his peers.

PREVIOUS PAGE: *The salmon fleet heads out from Bones Bay for a 6 p.m. opening, circa 1943.* John Mailer photo, courtesy National Archives of Canada, PA-145355

The fish come into the strait spread out from Cape Scott to Pine Island and "school up." Those that school up against the mainland shore head for Rivers Inlet. The rest head for the Fraser River and they come with the tide. The flood tide pours down Blackfish Sound and races through Blackney Pass and across Johnstone Strait. Blocked by the eastern shore of Vancouver Island it turns and rushes south. Dotted along this eastern shore opposite Blackney Pass the seiners lie close in, their engines idling, waiting for the invisible torrent and the fish it brings. Each skipper has his own gut feeling about where the fish will "hit the beach" and has positioned himself as advantageously as possible.

Experience and intuition have made some of these men masters at this skill but it is always an inexact science. Sometimes boats farther north have broken up the schools. Sometimes the fish come through a different pass into the strait and hit the beach farther south. Sometimes, getting the last of the net aboard, a skipper will see a school of salmon so large that they are finning on the surface swim right by and into the net of the next boat.

So the skipper and his crew are tense. They stand together on the dodger scanning the water and listening to the marine-band radio. The radio offers clues as to the fish's whereabouts, for the airwaves carry any number of cryptic messages that can sometimes be interpreted by a knowledgeable listener. At the stern the skiff is hanging off the back of the seine table, its bow a metre above the water.

No matter how young or old, how tired or hungover, how many weeks (or years) the crews have been at it, the thrill of these moments of anticipation never fades. This is the primitive thrill of the hunt: salmon fever. The adrenaline set coursing through each man's veins will carry him through a week of labour that continues almost non-stop until the closing on Thursday at 6 p.m. Searching for fish, all but the cook stand up there concentrating fiercely on the waters around them.

The radio blurts out a message that is carefully coded to make sense only to a select few listeners—the boats crewed by friends or family. Those up on the cabin roof stand transfixed. Although the message is garbled the voice can be identified. Snatching the binoculars the skipper searches the strait for the boat that has issued the message. He is searching for any signs of activity but there is none. His own boat has drifted some distance down the strait from the chosen tie-up spot by this point. He hands the glasses to a crewman and puts in the clutch. To avoid a puff of exhaust smoke they are still in idle. Back in position again they drift and wait.

The skipper is getting edgy now. His head swivels first ahead and then astern. He nudges the boat forward again. Nobody speaks. The skiffmen fasten the snaps on their oilskin jackets so that the metal suspender buttons on the bibs of their oilskin pants are covered. When all hell breaks loose and the two of them hurtle into the skiff from the pile of net on the seine table there must be no protuberances to catch in the webbing and sweep them down into the water to their death. The lead line on the bottom of a 12-metre net weighs nearly a tonne. It would take an hour to retrieve a body, never mind a survivor.

The crew's concentration is wavering now and they are fiddling with their clothing and blowing into their cold hands. No one wears gloves, for gloves, too, pose a danger. So their hands are cracked from exposure to icy salt water and wind. Their hard hats are held in place by the hoods of their jackets. Their heavy wool pants are two sizes too large and held up with suspenders, which makes them easier to abandon should their wearers end up in the water. Their hip waders are pulled up but not fastened, for in the event that they should fall overboard the waders will fill with water and must be jettisoned, like the pants, before they drag their wearers to the bottom. Every item of a fisherman's clothing, it appears, has the capacity to turn on him and drown him.

But then it happens. Perhaps it's a "jumper" that catches someone's eye, or finning on the surface of the water or simply a few ripples. Someone shouts, someone else points, and as if propelled by an explosion, four men, ignoring the ladder, leap from the roof of the galley onto the deck. The skipper hauls the wheel over, throws in the clutch and gives the boat full throttle. With the wheel hard over, the boat shudders as it picks up speed and arcs toward the shoreline. On deck the two skiffmen clamber over the mass of net and dive for the skiff. A deckhand loosens a couple of turns on the skiff line, holds the rest of it on a cleat and stares up at the skipper for directions. The boat has described a wide arc and is running down the beach as close as 15 or 20 metres from shore in the direction of a big cedar—the "tie-up tree." Now the skipper raises his arm as a signal for the deckhand, glances astern at the skiffmen, checks ahead and then drops his arm. "Let 'er go," he yells.

No military exercise is carried out with more precision. In seconds the skiff's painter is flung from the seiner, and the skiff, no longer held bow out of the water, drops, yanking off the bunt end of the net and bringing some of the lead line with it. The boat is running down the beach now with the net spewing off her seine table, the brass purse rings hammering off the stern in rapid succession. At the same time the purse line hums through the snatch block guided by a deckhand. To get a kink in the purse line would be disastrous to the set and possibly the net. Even stopping the boat wouldn't really help for the lead line would continue to pull the net off. The deckhand, cringing at the very thought, makes sure it doesn't happen. Two hundred metres and the boat swings away from the shore and describes another broad arc. Wham! One coil of purse line is gone and in seconds the deckhand is feeding the other side of it into the block at lightning speed. The skipper, now steering while facing backward, begins to ease up on the throttle as the cork line bounces off the table and follows along its buoyant path. One of the skiffmen, meanwhile, has slammed the oars into the oarlocks and is heaving his weight against them, driving the heavy skiff across the water and onto the shore. As it touches the beach the other leaps out, races over the rocks and sloshes through the tide pools carrying 60 metres of four-centimetre manila. He has reached the big cedar tie-up tree now. While he still has some slack he gets six or eight wraps around it.

For a few seconds the weight of the lead line slows the set but as the wall of net meets the flood tide the line sings with the strain and starts to slip centimetre by centimetre around the tree trunk. The man flings

The advent of the powered drum made it possible to make many more sets in a day.
Gordon Henschel photo

another wrap around it and the line holds. With two-thirds of the net in the water the seiner turns away from shore and starts to tow the net in an arc against the incoming tide. The boat, her engine straining against the current, holds this giant *U* open for the permitted 20 minutes. At the bow and stern, deckhands are slamming plungers deep into the water; their noise and the bubbles they make imitate a killer whale and frighten any escaping fish back into the net. Then it's time to close up. The skipper pulls the wheel over and noses toward the tie-up tree. Fifty metres or so from shore he sounds a piercing blast from the boat's air whistle; on shore the skiffmen are once more galvanized into action. Speed is of the essence and they know it. With the circle of the purse seine almost complete, most of the strain is off the beach line. One of the skiffmen lets it go and scrambles over the rocks with the line bunched in his arms. While his partner leans on the oars, he coils it up; as soon as they reach the seiner he flings the coil at a deckhand.

The source of every crewman's wealth is now hanging there in a purse that is open at the bottom. To secure it they must secure the purse line. No one needs to be told to move it. They are all once more working with speed and precision. The skiffmen, their boat secured to the cork line, race down the deck. Each one grabs an end of the purse line, throws three wraps around the winch and starts hauling it in, coiling it to perfection as he goes. Slowly the rings at the bottom of the net begin to close. When the purse is completely closed, the rings are hauled up above the surface, bringing the lead line up with them. The deckhands swing the boom over and slide the purse rings onto a steel bar, called a "hairpin" for its likeness to that object, and haul the rings and the lead line aboard.

Then they position themselves at the bow and stern and begin hauling in the cork line and piling it on deck. Left floating it will surround the boat and foul the propeller, leaving her drifting helplessly too near the shore for safety. Hauling in this net is slow, heavy work and has its own time pressure. They dare not slacken their pace until at least a third of the cork line is aboard. Once they have enough of it on deck they feed the beach line through the power block and start pulling in the net. It hangs over the seine table from the tip of the boom in a broad inverted *V*, raining seaweed and shredded jellyfish down the necks of those below. A hard hat helps, for occasionally it's not only jellyfish but a dogfish that crashes down from above—or, on rare occasions, the boom itself. As is the case in much of the seining process, everything is under shuddering strain, and before steel booms came in, the wooden ones occasionally snapped and fell on the unfortunates below. Experience helps to protect a fisherman's life but doesn't guarantee it.

But all being well, one man coils the lead line as the net comes in and on the opposite side another coils the cork line. Between them one or two men, the ones catching the brunt of the jellyfish, cast the net back and forth so that it is piled evenly on the seine table. Finally there is only the bunt end of the net left in the water. In it swarm the fish.

Deckhands now ready the brailer, a metre-wide dip net with a bag made of heavy-duty herring web. Like the seine net it has small brass rings around the bottom laced with a light chain. A length of line connected to the chain is pulled to close and latch the bag at the bottom. Too large and unwieldy to operate without mechanical assistance, the brailer is attached to a sheave on the boom so that it can be swung out over the swarming fish. A crew member holding the brailer's six-metre handle jams the net down vertically into the fish and, at his hand signal, the winchman tightens his wrap and pulls the brailer up and over the hatch, guided by a rope held by a deckhand. Another deckhand, universally known as "the asshole man," trips a rope that opens the bottom of the brailer and sends the whole load of fish gushing into the hold. This is the money coming aboard. As the fish are brailed, three crew members are continually pulling in the web, and finally the last of the net and the last of the fish are hauled over the rail and onto the deck.

And then the whole laborious process is repeated, again and again.

There are many sets when the brailer doesn't get used. These are the times when only 50 or 100 fish swirl around in the end of the seine net. But when the fish are there and the prices are high, one good set carries the crew all week. Several make the whole season a success.

If the fish are there, the tides are right, the crew is good and the net doesn't catch up on the bottom, it's possible to make set after set. But like Monty Python's instructions for playing the flute—"blow in one end and twiddle your fingers"—it isn't as simple as it appears. A myriad of things can go wrong. Misjudge the tide and get caught in an eddy and the net will swirl back in a horrendous tangle that takes an hour to straighten out. Meanwhile, unable to use the engine for fear of fouling the propeller, the boat will drift along the beach and sometimes have to be towed into deeper water by a friendly competitor—who might himself need a tow one day. Or perhaps the $25,000 net will catch on a pinnacle of rock, and those on the boat will watch in despair as the cork line bobs up and down indicating

Opposite: After most of the salmon have been brailed aboard, a deckhand wields a pugh to finish the job. The sooner the net was emptied, the sooner it could be set again. From the collection of the author

that below the surface the net is tearing itself to pieces. Seals and travelling orcas, if entrapped, will swim down and out, but sharks won't. Caught in the net they roll over and over, chafing a huge area of the cotton web, and then tangle themselves in it until it rips. Worst of all, a surge of strong tide can snap the beach line and tear the head off the skiffman tending it.

But barring major disasters, a boat could, with luck, fish all day and fill the hold and then, in the deepening darkness, set off to find the packer. Their packer may be an hour away, and there will be the lights of other seiners wavering on the black water when they arrive. Perhaps third in line, they wait. Their turn at last, they come alongside; crewmen pick up pughs—long handles with sharp metal points—and start loading their fish into the other boat one at a time. They are, by now, exhausted and soaking wet. They stand on the rail and individually stab, sort and toss fish after fish into the packer—4,000 fish each hour.

As they return to the fishing grounds they hose the decks down and pump out the bilges. Only then do the skipper, who has been at the wheel since dawn, and the crew, who have been working with speed and concentration all that time, fall into bed. They get three hours' sleep at best, for breakfast is over by five and the first set made at daylight. By Thursday at 6 p.m. it is over.

In the 1940s and '50s and '60s, on Thursday evenings in September, approaching Alert Bay by boat was like inadvertently sailing into the middle of a movie set. You could imagine the posters advertising the film: "Raw Emotion," "Two-Fisted Adventure" and "The Northern Frontier," they would have read. Everything conspired to make it so. No smoke machines were needed to provide atmosphere, for at this time of day, as if on cue, the fog rolled in from the mainland and drifted along the waterfront, making its hard edges romantic and mysterious. Clusters of masts swaying with brailers emerged from this swirling greyness; at their tops, company flags hung limp in the damp air. On the boats below, as if galvanized by a shouted command for action, the players, awkward in rubber coats and rolled-down waders, shouted and surged and stumbled in a frenzy of activity. Everywhere water gushed out of hoses, poured down from docks, flowed over decks, sprayed into holds and spurted out of scuppers. From radios the beat of country and western revived the flagging. And everywhere excitement kept exhaustion at bay.

Each boat waited its turn to unload its last consignment of fish. The decks were hosed down one last time and the bilges pumped. Then those of the crew who didn't have a home to go to walked up the road to the Rainbow Cafe. Its chow mein was a welcome break from fish and potatoes and peanut-butter sandwiches, but most managed only a few mouthfuls before they fell asleep face down on their plates.

By Saturday morning the catch had been tallied up and the crew could pick up three-quarters of their week's earnings. Before collecting his pay, however, Barrie performed a weekly ritual. He fished for Canfisco, but next door at the ABC Cannery there was a primitive shower in a plywood shack at the end of the dock. It ran off the boiler and had a limitless supply of hot water. Nobody seemed to mind who used it. Each week he bought a bar of soap and on Friday he stood under the gushing hot water until the soap was completely gone. It took approximately three hours.

CHAPTER 7

High Boats:
Ermalina — Louis Benedict (ABC)
Frank Ellis — Sam Hunt (BCP)
La Paloma — Henry Seawid (ABC's
high boat for the season to date)

PIONEER JOURNAL, JULY 26, 1956

The *May S.*, tethered to the bottom, lay motionless in the afternoon sunshine. David and Barrie dropped the skiff into the water and rowed ashore to pick blackberries for their supper. The small west-facing bay trapped the sun, and even this late in the summer there were berries, heavy with juice, on the mass of brambles. As they picked, the rich warble of a robin echoing across the water was the only sound. In the bay the *May S.* was perfectly reflected in her shining sheet of water.

I will save this afternoon at Mamalilaculla for a rainy day, thought Barrie, for this is a "pole in the bush" moment. He was 17 or 18 when he found the pole in the bush. Anchored off the beach at some unnamed spot, waiting for the opening, Barrie took the skiff ashore with no particular intention other than to poke around in the woods above the high-water mark. He used the salal bushes to haul himself up the bank and then pushed through the undergrowth on an aimless expedition. Suddenly the tangled wall of salal and salmonberry gave way to a small clearing. Lying there, almost obscured by the underbrush, was a huge totem pole, the great eyes and pursed lips staring at the sky. Barrie was transfixed. It's here in this ancient, remote and beautiful place that my spirit belongs, he thought. It was a profound reflection for a teenager.

The two men rowed back to the boat, their fingers purple with fruit stains. The *May S.* was aglow in the light of the sunset out in Blackfish Sound. Sunsets on the water are extravaganzas. They start gently with shades of pale green mingling with the still-blue sky, then become orange and red. The few clouds in the sky turn purple, then black, and their lower edges blaze with gold. As the air cools a layer of fiery mist lies on the surface of the water. The whole thing is too much; there is no restraint. David squinted into all this glory. "There's all these things," he said. "There's no steelhead in the Vancouver Island rivers any more. Pretty soon there'll be no more salmon. There's so many pigs feeding at the trough—different pigs with different interests." He slipped the eye of the skiff's painter to the hook of the boom, and together he and Barrie hauled the little boat out of the water and onto the aft end of the *May S.*

"Time for a cigarette," he said.

They sat in the deepening darkness watching the blazing circle that was the sun drop slowly into the sea.

"Pretty peaceful, eh?" David said, and then he chuckled. "Wasn't always so peaceful around here, was it? Not when we were pulling in the fish. They'd close the Skeena and the Nass and then the Fraser and everybody would show up in Johnstone Strait."

"Everybody would sleep on Friday," Barrie said, "and then on Saturday they'd get a drag and hit the Nimpkish. And then the loggers would all come over. Water taxis just shovelling them into the Bay."

"And then, of course, the Bella Bella boys would be in town." (The "Bella Bella boys" and the Kwakwaka'wakw people of Alert Bay had been feuding for 200 years. It's only in recent times that the participants have halted their strife.)

"Somebody would start a fight and in no time the whole place was upside down."

"Somebody would smash a beer glass and shove it in somebody else's face," said David.

"Followed by a fist," said Barrie. "I used to go around the corner where the waiters' jackets hung and every once in awhile I'd take a look out. There'd be chairs going through the windows—and people going through the windows."

If Barrie's mother had had any conception of how different Alert Bay was from Victoria, she would, no doubt, have sent urgent messages for him to come home. But immersed in her job and the occupying task of looking after a drunk, she was unaware of the realities of life in Alert Bay. And Barrie, now part of this rowdy village, had no intention of letting her in on the truth.

Drinking, which had been pretty well confined to Saturday night, often spilled over from the weekend into the week—with predictable results. The *Kitgora*, for example, stemming the strong tide that raced past the docks at Alert Bay, was overtaken by a seiner with a BC Packers buff-coloured cabin. Her skipper's distance judgment (among other things) was impaired. He flew by the *Kitgora* at full speed and attempted to land at the dock ahead. It was not a successful attempt. With a splintering crash he struck the dock head on, and only the fact that he was holding tight to the wheel prevented him from being knocked down by the impact. The fat man standing on the stern was not so fortunate; he was hurled off into the water. By the time Steve got abeam of the boat there were a couple of men on the after deck flailing about in the water with a pike pole. This less than perfect landing had done nothing to dissipate the skipper's good humour, however. As the *Kitgora* passed him he grinned at Steve and took one hand off the wheel to wave unsteadily. "Not my boat," he yelled.

Another day, having left the *Kitgora* to fish on another boat, Barrie found himself with a skipper much less responsible than Steve Warren. Late one afternoon, despite being in an advanced state of intoxication, this man managed to get his boat fuelled up, backed her out into the bay and headed her down the strait toward his destination, Campbell River. As Alert Bay receded in the distance this man was struck by the fact that the Nimpkish Hotel was receding in the distance, as well. This thought caused him pain. He pulled the wheel hard over and the seiner heeled on her side and turned around.

"Where you goin' now?" Barrie and another young crew member

were standing up behind the dodger with the captain and this sudden change of direction caught them off guard.

"Goin' back to the Bay. Just goin' to run her up on the beach by the Nimpkish."

The two startled crew members tried to reason their way out of this dilemma but when that didn't work they resorted to force. Even in the casual hierarchy of the fishing fleet, crew members didn't contradict the skipper, but now these two lanky adolescents threw that stricture to the winds and wrestled the wheel away from their protesting captain. They persuaded him to retire to the galley where he and the cook slumped over the table in an unconscious tangle. There they remained, unconcerned by the fact that their boat was being guided through the gathering darkness by two teenagers whose aids to navigation were the lights of the boats ahead. They followed those lights all the way to Campbell River.

The problem, of course, was one of excess, or rather the perception of excess, which is the same thing. In every direction there were trees as far as the eye could see, from the mountaintops to the waterline. They crowded down to within a metre or so of the high-tide mark, their branches forming a line above the tide line as straight as a hedge. Under the surface of the water there appeared to be an inexhaustible supply of fish, for each year, as technology improved, the fishermen were hauling more and more out of the depths. The work itself was excessive; to slave all week with virtually no sleep, as the fishermen did, or to heave a 30-kilogram Pioneer chainsaw up and down a sidehill for eight hours, as the fallers did, would kill most city dwellers. And sometimes the money was excessive. For the fishermen it came in like a lottery; one year they were just making their costs and the next they were lurching in so heavily loaded that their decks were awash. Even the children in the village were beneficiaries of this excess. In the backyards of their houses every Tonka toy known to man lay rusting in the rain.

In 1951 Area 12, which was the fishery from Adams River to the north end of Vancouver Island and over to the mainland, produced more fish than any other area on the coast. Some of its fishermen went home to wives and families and put money in the bank. The Italians and the Yugoslavs saved their earnings and bought investment properties in Vancouver. The rest swarmed along the gravel road that curved along Alert Bay's waterfront and ended up in the Nimpkish Hotel.

There is a scene in the otherwise forgettable movie *The Perfect Storm* in which Bugsey, a fisherman off the swordfish boat *Andrea Gail*, tries to pick up a woman in a bar. The *Andrea Gail* has just come in after weeks at sea; next morning she'll be heading out to the fishing grounds again. And Bugsey fears that he alone, of all the crew, will not get laid during this brief turnaround. In some desperation he strikes up a conversation with a buxom world-weary woman on the next bar stool but his clumsy advances are rebuffed. It's a touching scene—a well-acted Hollywood version of real life. The real thing occurred every Saturday night in the Nimpkish Hotel. It was far from touching. The women in the Nimpkish weren't world weary; they were giggling Native girls escaping boredom. They were searching for attention, glamour and excitement. None of these things, of course, was available at the Nimpkish.

The events of the evening had a sequence of their own. The man who hit on a girl early in the evening was rejected. "Get your hands off me you goddamned chicken ass. I know what you want," was the more or less amiable response. But later in the evening, plied with more beer than she could handle, the same girl would leave, sometimes with several men, bound for some dark spot in the shrubbery or on the beach. Sometimes, dazed with alcohol, the girls collapsed at the side of the road on their way home and were picked up by a passing fisherman and packed off to his boat.

"In Alert Bay it was a shrewd man who knew his own father," Don Pepper said.

"Looking back on it, it was awful," Barrie said.

Although Rod Hourston was intimately connected to the fishing industry he wasn't a part of these brouhahas. He graduated in 1949 with a master's degree and joined a new branch of the Department of

Fisheries called Fish Culture Development. It was a bioengineering group concerned with stream clearance and improvement, hatcheries and fishways, and it was established at a time of unparalleled industrial development. Pulp mills were being built, vast hydroelectric projects planned, whole towns, like Kemano, springing up to service resource industries. And the forest industry, with a workforce now released from the military, continued to boom. The BC government encouraged industry's growth with a policy of what Stuart Leggatt, the judge who was to become the commissioner of an inquiry into salmon farming, later called "sympathetic administration"; in other words, business interests took precedence over the protection of resources.

No one found fault with this policy, for 10 years of depression and six of war had created a population starved for some of life's small luxuries. All the industrial effort that had fuelled the war was now focussed on creating prosperity. There were jobs for everyone, and for the first time in some people's memory there was such a thing as discretionary income. BC Hydro advertised regularly in Alert Bay's little newspaper; perky wasp-waisted housewives stood beside gleaming white stoves and washing machines and extolled the unbelievable ease that electricity conferred. Today's society—one in which the possession of microwave ovens, stereo systems, TVs and weed-whackers are considered a God-given right—can have no conception of the austerity of that earlier time. So the booming economy was welcomed with open arms. That this prosperity came at a price wasn't immediately apparent to the general population—and its leaders would be safely out of office before the bills came due.

In the 1940s and '50s fisheries management was still pretty basic. Up until that time there seemed no need to do research in this vast silent wilderness that was the BC coast. Only a scattering of people inhabited the twisted kilometres of coastline, and in those post-war days they were incapable of leaving any significant mark on their surroundings. Before the age of bulldozers and chainsaws, loggers could only gnaw at the edges of the limitless forest; before electronics

W.R. (Rod) Hourston, director of the federal fisheries department's Pacific Region. He was charged with protecting the fisheries at a time when industrial growth was the province's top priority.
W.R. Hourston photo

and hydraulics, fishermen pulling in their nets by hand could only make three or four sets in a day.

With two exceptions there were no biologists involved in the management of the salmon; the exceptions were the scientists working for the International Pacific Salmon Fisheries Commission and those employed by the Fisheries Research Board of Canada. The former were involved in the restoration and enhancement of the Fraser River runs following the construction of the Hells Gate fishways in 1945. The latter were working exclusively on the Skeena River where a slide in 1951 had created an emergency situation.

In the waters around Alert Bay the employees of the fisheries department were field people—fisheries officers, patrolmen and guardians—who covered the area by boat. The boats of the day were small and slow by today's standards and there were not enough of them to "cover the waterfront." Realistically, all the DoF's employees could hope to do was to stop any visible abuses in the saltwater environment. And the fines for the transgressors who were caught were so small that they robbed the department of any real clout. The investigations of the spawning grounds themselves were carried out by fisheries officers and guardians who regularly slogged up streams through the underbrush and made the best estimate they could of the number of salmon spawning there.

The establishment of the Fish Culture Development Branch proved to be a fortuitous event. Rod Hourston and the other biologists and engineers hired by this branch found that their work

Ian Todd began his career as a 19-year-old biology student studying the Native fishery at the mouth of the Nimpkish River. Ian Todd photo

became increasingly important—and increasingly difficult—as massive hydroelectric projects began to take shape in BC. Their most pressing assignment became the task of protecting fish from the effects of the post-war boom. For suddenly the DoF found itself in an impossible race with the industrial expansion that was bursting out everywhere in the province. Suddenly logging was trashing the spawning streams, dams were blocking access and manufacturing was polluting the waters.

Ian Todd was one of many new employees involved in this work. In 1955 he was a 19-year-old UBC biology student. Built like a Mack truck and happiest in the outdoors, he was pleased to find himself spending the summer on a little island at the mouth of the Nimpkish River. He lived in a fisheries guardian's shack there and his job was to study the Native food fishery. "When the Indians came across with their beach seine, I essentially bought fish from them, tagged them and released them into the river," he said. After awhile someone supplied him with a small boat and a five-horsepower motor so that he could get across Broughton Strait to the bright lights of Alert Bay.

There was a pressing reason for Todd's presence there on that soggy little island in the middle of nowhere. Among the major hydroelectric initiatives being considered for Vancouver Island rivers was the damming of the Nimpkish. The DoF was scrambling to collect data and assess the impact it would have on this prime spawning area. Whether the DoF's efforts had any bearing on the final decision is debatable, but the plan was shelved and a year or two later, when Ian Todd graduated, he returned to the area as a biologist. He was part of a group studying the Nimpkish salmon in detail for the first time in history. In the scheme of things the status of this group was hardly impressive.

"We came up to Alert Bay on the Union Steamship," Ian said, "and took a water taxi to Beaver Cove. We bummed a ride on the Canfor speeder to Nimpkish Lake, took their tug to the top of the lake and another speeder to Woss Camp. We were allowed to board at the camp and we hiked in and did our work from there."

The fact that the DoF was dependent on the logging companies for transportation and accommodation shouldn't be lost. Like poor relations these scientists were tolerated by their hosts, but there was no doubt where the power lay. "In those days logging wasn't trying to conserve a resource—any resource," Todd said.

By the 1970s this was apparent to anyone who cared to think about it and people were becoming alarmed. A January 1973 issue of the *Vancouver Sun* contained a picture of the Cameron River watershed on its front page; the river ran through a valley completely shorn of trees. The Sierra Club held meetings at which a copy of Crown Zellerbach's annual report for 1971 was circulated. Full-page, full-colour photos showed logging right up to a stream bank and the stream itself virtually invisible under the debris. The fishermen of Tofino formed the Save Our Salmon Association and went to Victoria to protest the damage done to the Indian River. "It's not just the Indian River," said association member Doug Arnet. "It's every stream on the west coast of Vancouver Island. It's an incredible mess and neither the provincial forestry nor the federal fisheries people have shown any concern about it."

Rod Hourston, of course, was trying to conserve the resource, and his battle with the powerful forestry lobby was never-ending and frequently disheartening. "Problems with logging?" he said. "Don't get me started." Above the streams the clear-cut logging had tragic consequences. It changed the runoff dramatically. "Vancouver Island was hit pretty hard, but in those days the central coast hadn't been logged too much. But then they started logging in Rivers Inlet. I flew over it. Instead of a green canopy there were naked brown river valleys. And logging roads—logging roads are the most damaging thing. The mud and gravel run off into the culverts and then into the streams and it's just a mess. I've seen so many messes," he said wearily.

Occasionally circumstances favoured the DoF. There was the plan, for example, put forth by the logging companies on the north coast to drive logs down the Nass River. "We were battling it," said Hourston, "but they won out and started sending logs down the river. And they tried to drive the Skeena the same way. But it didn't work. The West Coast timber is a hell of a lot different than those little pulp logs in the east. They lost so many logs by stranding that they finally gave up." For the fisheries it was a victory by default, of course, and an almost negligible one considering the scope of the ongoing destruction.

One of the most dramatic examples of this destruction occurred in 1957. Three years before, an outbreak of the blackheaded budworm had infected the forests of Alaska and had begun to spread down the coast to BC. The Forest Service watched this situation with alarm and decided to initiate an aerial spraying program. DDT is lethal to fish—that much was known even then. Furthermore, the spraying program would be carried out in the spring, the very time the salmon smolts would be in the rivers. So Rod Hourston, then the DoF's chief biologist, and three other biologists flew to Port Hardy and prepared to do battle on behalf of the North Island salmon. At the very least they hoped to negotiate "no-fly zones" to protect the most vulnerable sites. At the Port Hardy airport they met with BC's chief forester and the chief pilot from the aerial spraying company to discuss a possible flight plan that would minimize damage to the fish.

Avengers chartered by the provincial forestry department sprayed North Island watersheds in 1957, laying waste to salmon fry. Photo courtesy UBC Library Rare Books & Special Collections, MacMillan Bloedel Photographs

Yet even as they spoke they heard the roar of a Grumman Avenger taking off. The bugs were out and there was a break in the weather; the spraying program had begun. For a week or two the Avengers flew over a 60,000-hectare area from Blinkhorn Peninsula to Port Hardy, drenching the region with a mixture of diesel oil and DDT.

In order to measure the impact of the spray the fisheries department had constructed pens in one of the local rivers and stocked them with coho smolts. Just one pass of the planes and every one of these tiny fish was floating on the surface of the water, victims of the deadly pesticide that had rained down upon them. On the banks of the Nimpkish, Cluxewe and Quatse rivers, dead and dying fish lay in piles and blinded fish struggled in the water. According to local reports 35 per cent of the pink and chum fry were destroyed and 85 per cent of the coho smolts. The sockeye, still in the waters of Nimpkish Lake where the pesticide was more diluted, fared better. No estimate could be made of the secondary effect of poisoning the fish's food sources.

Environmentalists were thin on the ground in the late 1950s. Surrounded as they were with pristine wilderness, the local inhabitants couldn't envision its destruction. The account of the incident in *The History of Alert Bay*—an incident that would have present-day activists gathering on the lawn of the legislature—is muted and naive. "As this district depends to a great extent on the fishing industry," the book says, "all concerned watched with immense interest this scientific undertaking. It is to be hoped that the final results will show no detriment worth speaking of to this vital industry."

A.J. Whitmore took a less sanguine view. Whitmore was the chief supervisor of fisheries and he was outraged. In Vancouver he called executives from the timber companies into his office and did a "Nikita Khrushchev," as Ian Todd put it, pounding his desk with his fist. "He was a very powerful figure," said Ian, "and he really gave it to them." They left the office shaken but otherwise unharmed, for the maximum fine imposed by the Fisheries Act at that time was $25,000, a sum the forest companies could consider the cost of doing business. Nonetheless, the program was a clear violation of Section 33 of the Fisheries Act, which prohibits "the introduction of any substance in the water that is deleterious to fish." On this basis the spray program was abandoned.

"But it was always a battle," said Rod Hourston. "If you weren't trying to save the fish from this kind of thing, you were trying to cope with the technological advances in the harvesting techniques. At one time they came out with a monofilament net. It was practically invisible. It was so effective that there wouldn't have been a fish left. We finally managed to ban it. Even the fishermen got upset. They said, 'We're too efficient.'"

"It didn't happen all at once," said Ian Todd. "When they got the power block they were making eight or 10 sets a day instead of three

ALL
CAPE HENRY
CAPE GEORGE

ELVA M II

or four. And then they started using drums and they were making 20 or 30 sets. It was a whole bunch of little things—and they all began to make a difference."

One of the problems was that the "seine-boat boom" that had started in the 1920s was finally coming back to haunt the industry. By continuing to supply the money and the encouragement to buy new boats, the fishing companies had created a fleet much larger than the resource could absorb. A second problem, of course, was that this excessive number of boats had become extremely efficient. The Alaska government, more far-seeing than the Canadian one, banned the use of drums in the seining industry to protect its resource. But below the border the Canadian government favoured the interests of capitalism and permitted their use.

Large-scale logging and a bigger, more efficient fleet were decimating the salmon fishery. The massive diking that took place on the Fraser after the 1948 flood may have had an effect, too. These dikes cut off the sloughs and meandering little streams that supported salmon stocks.

In any event, suddenly, in the mid 1960s, there were no salmon. The fisheries did a major spawning estimate on the Fraser and found the stocks had dropped to 165,000—less than a quarter of former levels. Shorter openings were not enough to stem the decline. At one point the whole fishery was closed down.

Now the fishermen of Alert Bay had to depend on the herring fishery to keep them afloat financially. The Cook family had fished herring from the very beginning; in the winter they tied up their seiners and the fishing company leased them larger boats, for the herring fishery required a bigger vessel—a boat that was as much a packer as a seiner. And then they switched nets—for the nets were bigger, too—and ranged the coast from the Gulf Islands to the Nass. There were lots of herring.

PREVIOUS PAGE: *Typical seiners of the boom years of the 1920s.* Photo courtesy UBC Library Rare Books & Special Collections, Fisherman Publishing Society

Darrell Huson, lost when the Amaryllis, *a herring packer he was skippering, sank on February 7, 1948.* From the collection of David Huson

"You'd be rushing around a big herring ball," Steve Warren said, "trying to find a place to set. So many herring they filled the boats plumb full. Most of the time those boats had no freeboard—the decks were awash. Lots of those herring boats, if you walked from one side to the other it leaned that way. It wouldn't take much to sink it. Or if there was a leak someplace, down it would go. Happened many, many times."

The boats were overloaded because this was the "reduction" fishery; the fish were processed into fish-meal fertilizer and oil, not food. Consequently the prices paid for them were low and it took a lot of fish to make any money. "When I was out fishing with my dad and Steve," George Cook said, "for 1,000 ton we'd make $100 a man."

And so they continued to cram their boats with herring. On a cold February day in 1948 in Rivers Inlet, David Huson's uncle, Darrell, was doing this. He was running the BC Packers' boat *Amaryllis*, and a pool of boats was brailing tonnes of herring into her hold. "Load her up," said a jubilant crewman. "We've all got kids at home."

By the time they got their net aboard and started out for Namu, the *Amaryllis* was staggering under a 100-tonne load and answering sluggishly to the helm. "But she'll be okay to Namu," Darrell Huson said.

It took a lot of herring to make a decent payday in the herring 'reduction' fishery, and it was hard for a skipper to resist the temptation to overload his boat. David Huson photo

"It's all inside." At Namu, though, the reduction plant was swamped with herring. Incapable of taking any more fish they sent the *Amaryllis* south to Steveston. She retraced her route down Fitz Hugh Sound, passed Cape Calvert and entered the open water of Queen Charlotte

Strait where she met a winter gale. In a blinding snowstorm she waded into mountainous seas.

An experienced skipper knows when his boat doesn't "feel right." The *Amaryllis* felt dangerously wrong. Darrell Huson did what he could; he sent the mate out to the wave-washed afterdeck to make sure the hatch cover was secure, and he radioed a following seiner, the *Bligh Island*, to stand by in case of trouble. For two hours both boats continued to fight their way south, then the wallowing *Amaryllis* began to sink. Somewhere off the light at Scarlet Point she rolled and didn't recover, and her crew, their lifeboat smashed, took to the water. Though her searchlight was blurred by falling snow, the *Bligh Island* managed to find two crew members. The *Amaryllis*, however, along with her captain and her tonnes of herring, went to the bottom. In Alert Bay Darrell's mother, sensing tragedy, went down to the dock and waited for hours in the dark and the wind and the rain.

The price of herring required huge catches to make a living, and to increase them the ever-ingenious fishermen began using lights to attract the fish. These were mercury vapour lamps, the kind used as street lights, and they were mounted on a curved metal pole that swung out over the ship's side and glared into the sea.

"The fishermen came out with those lights—those pit-lamps," said Rod Hourston. "That was just a leap of efficiency. You could see the fish just building and building under the boat and then, in the morning, out go the nets and then poof." This pit-lamping was devastatingly effective. The herring stocks diminished year by year and finally collapsed. "We closed the fishery because of this over-harvesting," said Hourston. "It was closed for six years but fortunately the herring did come back."

Meanwhile back in Ottawa, a very long way from Johnstone Strait, ministers were invited to good lunches and the lobbyists, their hosts, got the results they wanted. Sometimes all that was required was inaction.

"Wasn't so many of us a couple of hundred years ago," said David. "People just catchin' enough to eat. Nobody thinkin' of selling fish.

Phenomenal returns from the herring roe fishery prompted David to buy a second boat, the North Isle. David Huson photo

Then they started that saltery and they started catchin' fish by the ton. They didn't want anything but the red springs for saltin' so they just threw the rest away. They got engines then and stuff like that. No more rowin' around in an open boat."

"It was drum seiners that made a real difference," said Barrie. "When they first started showing up with seine drums we never figured it would work with that big gap open alongside the hull when they were hauling the net in. We thought they'd lose half their catch. Then they had to hang their net different, too. I always wondered what they'd do if they got a 'roll-up'. Can you imagine trying to pull that apart with your bare fingers while hanging over the side?"

"Tell me about it. I've done it lots of times. Don't forget I've had a drum for over 30 years now."

"It was that and the electronics," said Barrie.

"It was coloured sonar that changed the whole thing," said David. "We all had those old black and white sonars, but when I was in Nanaimo on the *North Isle* one time a guy came down on the dock and said he had this coloured sonar he wanted to try on a boat. He offered it to me. And I just laughed. I was going herrin' fishin'. I had to be out there the next day. So I just laughed, eh. It probably would have changed everything around at my end of it. Next thing you know we were all buyin' coloured sonars. And I mean—the difference in the fishin'! Think about it. Tommy Hunt had a piano wire. With colour you can tell the density of the fish. It's absolutely amazin' to watch. Just like lookin' at the bottom on TV."

"Which is why they had all those squid boats out there when that plane crashed in the water in California."

"You bet. They could pick up the outline on the bottom."

"And we thought that when the first boats got radar it was really something. They could make a set in the dark and we couldn't."

"You had to be able to see the tree on the beach."

"Sure, but still they could set in the fog. We had to wait 'til after breakfast. A real heavy day for us was nine sets. How many sets can you make now with a drum seiner?"

"Twenty," said David. He picked up his coffee mug. "Before it gets too dark, I got to check that anchor. Make sure it's holding."

CHAPTER 8

High Boats:
Barclay Sound — Jack Joliffe (BCP)
Wayas — Archie Robertson (ABC)

PIONEER JOURNAL, OCTOBER 4, 1950

Charlie Pepper, his wife and their gaggle of small children stepped off the Union boat in Alert Bay in March 1941. World War II hadn't yet brought the full employment that wars create, and the legacy of the dirty thirties was still apparent. The mill at Englewood closed that same year, and in the city good jobs were still scarce. So Charlie, a stationary engineer, was happy to have become the successful applicant for the job of engineer in the heating plant at the residential school in Alert Bay. The Anglican Church wasn't a munificent employer but it was a reliable one. Charlie's modest salary would be paid without fail, which was important to a man who would eventually have 10 children to provide for.

And Charlie Pepper was an indefatigable provider. His job at the school was just his day job. He showed movies on Saturday nights, he started a bottle depot and he did whatever else he could find to make a dollar. Eventually his tireless efforts allowed him to build Alert Bay's first movie theatre.

Don, the oldest in the Pepper family, was six when his family came to Alert Bay. Like his father, Don was a worker. At school he inhaled 12 years of the required information in 10, and then he joined the real world and did what every other young man in Alert Bay did: he went fishing.

He learned the trade from the ground up, which meant that in an immense, echoing net loft at Bones Bay he helped make the seine net itself. Measurements in fathoms and other more esoteric calculations were marked on the floor there, and the net was designed for the type of boat that would use it. In 1953 they were still cotton nets with Spanish cork floats, and when the fishery opened they set them five or six times a day and hauled the nets and their loads of fish in by hand. "You need young, stupid kids for work like that," Don said. Don may have been young but he certainly wasn't stupid. He was a big kid, too—six-foot-three—and a hard worker, so skippers were always ready to take him as crew. If they could put up with his endless questions, that is.

"We'd go to Blinkhorn," said Don. "That was the traditional spot for the Alert Bay fishermen. Alert Bay fishermen knew two things: Blinkhorn and Double Bay. It was just a big circuit. On the start of the flood you'd fish Blinkhorn. A big surge would come across. The best sets were at high water. And then when the tide went bad you'd either go through the Blowhole or through Blackney Pass into Double Bay, and that was an ebb tide setting."

In 1956 Don fished with Vern Skogan. They left the confines of Johnstone Strait and went north. "We went to Namu. We went across the sound. Wow! It was a big thing. In those days I never saw a chart; there was no sounder, no radar on the boat."

At Namu they fished for pinks, and Don made a lot of money that year and in the succeeding years. He spent it on a university education. He wasn't the first to finance his way through university by catching fish; there are a lot of Vancouver's professionals who owe their corner office with a view to salmon. But Don Pepper was probably the first to attend the London School of Economics and end up with a Ph.D.

It was the perfect time to start a fishing career. Technology was changing everything. In September 1955 the first power block appeared in Alert Bay. It hung from the end of the boom and the net was drawn through it. The first ones were rope driven; "It was just a nightmare to get them to work," Don said. But gradually they improved, and then instead of hauling in the net the crew had only to

lay it out as it came through the block. With a power block, on a good day, a boat could make nine sets, so despite the usual naysayers it was only a year or two before every boat in the village had one. They needed to fish more efficiently, for radio telephones had arrived, which changed everything. Their ears glued to the marine band, the fishermen now knew where the best fishing was. "They'd think, well it might be just as well for me to move into that area," Steve Warren said. "And so instead of 40 boats in Johnstone Strait there'd be 80. So then it was local knowledge that was important. To get in the best position the local boats anchored out the weekend before the opening. Those poor guys coming up from the Fraser River, they didn't know where the hell to go. And we didn't tell them either."

And then the *Blithe Spirit* arrived in Alert Bay, and the promise of World War II technology arrived with her. The *Blithe Spirit* belonged to Jim Spilsbury, the pioneer manufacturer of marine radio telephones on the coast. She was a floating electronics laboratory and an advertisement for the wonders that were to come. She carried prototypes of "fathometers" (depth sounders), automatic pilots and radar. "It's been a long pull," Jim Spilsbury said, "but we expect that all the bugs will be worked out by the end of the year, and all being well these models should be in production next spring. Very soon I expect to see a radar set on every fishing vessel on the coast."

That's what happened, of course. Radar made it possible to set before daylight or in thick fog. Their appetite for technology whetted, boats then began to get walkie-talkies as well, which kept their secrets more secure than the radio telephone did. And by the late 1950s they had sonar. Tommy Hunt's piano wire was a thing of the past. Then coloured sonar became available. And by the 1960s the power block was being phased out. The fishing fleet was hauling its nets in on a big

Opposite: Aviation and electronics pioneer Jim Spilsbury at the helm of the Blithe Spirit, *bristling with navigation and communications gear. Spilsbury predicted, accurately, that before long he would 'see a radar set on every fishing vessel on the coast.'* Photo courtesy UBC Library Rare Books & Special Collections, A.J. Spilsbury Photographs

drum that sat at the stern where the turntable had been. At first everyone said that drums wouldn't work, but the Martinolich family made them work and then everyone wanted one. David Huson's father was one of the first to buy a drum. Steve Warren never did. "I was too cheap," he said.

But the ones who bought them were making 20 sets a day. Fish were pouring into holds and into the canneries, and money was pouring into the fishermen's pockets.

One of those pockets was Don Pepper's. Each summer he was out there catching the fish that would finance his next year at university. Alert as a prairie dog, he identified the outstanding skippers who fished out of his home community and got himself a job with one of them.

"In the straits there were three or four guys who were considered masters," he said. "One was Louis Benedict (the Fox) on the *Ermalina*. Pretty boat. Always been lucky. Fishermen believe in luck, you know. Then there was Freddie Joliffe; he was good. Gilbert Cook was the dog-salmon man and Reg Cook was the master of the straits."

Don fished with Freddie Joliffe. "Freddie really knew Blinkhorn. On a big tide there would be two surges, so if you were in the right spot you could really do well. That was the key to Blinkhorn that a lot of people never understood. But Freddie did. I'd say, 'What are we doin' stayin' here? All the other boats have left. Why—'"

Freddie, not one given to explanations, cut him off right there. "You want to catch fish, don't you?" he said.

In the absence of instruction Don made copious notes and analyzed his skipper's every move. "Fishermen are so superstitious," he commented. "They said, 'Freddie's lucky.' Well, I fished with Freddie for five years and I know where the luck came from: hard work. And when the fish came we were ready. Because we were just killers."

He kept on processing information, listening to men who had fished the coast all their lives. "Cohos and pinks are easy to catch—piece of cake. Tow for half an hour and you've got 'em, every one. They're just poking along. Not with sockeye. They're *activated*. When you're fishing sockeye—same place, same tide—12 minutes 'cause

Fishing with highliners provided Don Pepper with the money to put himself through university. He emerged with a PhD.
Don Pepper photo

they're in and out." On big tides, he learned, the fish were moving faster than on small tides. Outside Growler Cove at the Pig Ranch it was flat as a pancake; no tide in there. But on a big tide a back eddy would form and you could set there. You never got sockeye, though, never got sockeye on that north shore. The sockeye were always over on the Vancouver Island side.

Don kept his eye on the other master fishermen, too. "From Izumi rock to Robson Bight was Louis Benedict's domain," he said. "He knew how to fish Baronet Pass—pioneered it there. It was called the Bank because it was only open certain times of the day. It was a spooky place to fish. So much tide. You needed excellent gear—an excellent beach man, beach lines, boat. Lots of power. I've set there and the whole net completely sank. The undertow just yanked it under. Nothing showing but the purse lines."

These palmy days, however, were coming to an end. High-tech logging and high-tech fishing had taken their toll on the salmon and there were no signs of a recovery despite the closures. Nineteen sixty-six and '67 were bad years, and by 1969 there were no fish to be caught. Don Pepper left the fisheries that year. He had made exactly $712.

"Some people were just trying to fish long enough to qualify for unemployment insurance," said George Cook, "but that wasn't earnin' a livin'. It was just keeping the wolf from the door." George had skippered the *Winifred C.*, but in 1958 he contracted tuberculosis and by

the time he got out of hospital the closures were longer than ever and the fish fewer. "I sold out," he said. "I just couldn't make it. I just moved out of it altogether. You pay $10,000 insurance for the boat and wharfage and things like that. There just wasn't enough fishing time to pay those bills."

The *Winifred C.* was the first of the Cook's fleet of six to go. In time all but one of the Cook sons—and their sons—would sell their boats. "There's Reg," said Steve Warren. "He's the only one with a licence now."

Fortunately Stephen Cook didn't live to see his family fleet disbanded. He died in 1956, and the residents of Alert Bay filled the little white Anglican church and overflowed onto the steps and into the street. His pallbearers included four cannery managers.

When the fishing was poor, David Huson took odd jobs to make ends meet. At 18 he filled in as deckhand on the *Lady Rose*. The job paid eight dollars a day and it was a long day. "Four a.m., high water, we're in Port McNeill getting the cars on. Down to Kelsey Bay. Eleven p.m. we're back in Port McNeill again—just pulling in."

Then in Muchalat Inlet the whole town of Gold River was emerging from the forest, so he drove down the island to look for a construction job. The road started at Kelsey Bay and at Campbell River it turned westward. It was a road so narrow that the alders on either edge swiped the sides of the car. On the rare occasions when he encountered another car, or a wheezed-out Vancouver Island Coachlines bus hammering and shuddering through the rain-filled potholes, each vehicle climbed right into those alders to squeeze by the other. He had scarcely started this new job when it all stopped. "All the crew went on strike," he said. "The cookhouse locked everybody out; 1,300 men all tryin' to get in the same cafe. They finally got it straightened out. It turned out to be the milk machines. They had two, and when they made one of them chocolate milk that's all there was. Just something to get a few extra days off, I guess." It was different from fishing.

Opposite: The Winifred C *was the second vessel built for the Cooks, and the first to be sold. Eventually Stephen Cook's heirs would sell all but one of the family's fleet of six boats.* Photo courtesy UBC Library Rare Books & Special Collections, Fisherman Publishing Society

By the early 1970s he was at Quatsino working for Frank Hole, towing logs with the *Walter M.* "There was just the two of us on the boat," he said. "We weren't making all that much, maybe $10 or $12 an hour, but we worked 24 hours a day. Never stopped—only to fuel up. Frank was pretty good; he'd come out in his boat and tie up beside us. He brought us grub and took over while we got some sleep."

There aren't many logs coming out of the woods in January, so one year David leased the *Walter M.* and took her around Cape Scott, up Kingcome Inlet and into Wakeman Sound to fish for herring. He and his father had made a deal to fish for a Japanese company, but when David came alongside the *May S.* in Wakeman Sound his father had bad news.

"All the packers have left us," he said.

The Japanese company hadn't made an agreement with the fishermen's union and the packers refused to pack for them.

"My dad had hired one man; I had hired six. We figured we had to salvage somethin' out of the season. So we fished together and packed 35 tons each back to town."

In Vancouver BC Packers was ready to buy their fish, but on the company's terms. This meant that virtually all their herring money would go to pay off the debt that Spence Huson carried on BC Packers' books. Like most fishermen (and like farmers, for that matter) Spence needed a loan each year to finance the many large expenses that fishing entails. He borrowed that money—for a new net, for mechanical and electrical equipment, or for maintenance and repairs—from the company, and his lender felt justified in dictating the terms of repayment.

Unlike the Cooks' relationship with the Canadian Fishing Co., the Huson's relationship with BC Packers was often acrimonious. Now it flared into an angry confrontation as Spence Huson tried to get some cash out of the deal and the company maintained its position that the money must all go toward the debt. Spence and the company's managers parted in a rage. It turned out to be the best thing that could have happened as far as David Huson was concerned, for the Husons went to Bob Bingham with their problem.

Bingham had been manager of Babcock Fishing and had recently formed his own company. David had known him for years. They sold him their fish and then went north and brought him down four more loads.

"How about fishing salmon for me this summer?" their benefactor asked.

"I would if I had a boat," David said. "This is a BC Packers boat."

"If I find you a boat will you fish for me?"

It was the chance David had been waiting for all his life.

Two thousand years ago a Native fisherman and his *klootch* paddled out into the tide and trolled for the salmon that would keep them from starving in the coming winter. Their canoe, their paddles, their lines and hooks and sinkers had been made by themselves and the people of their village. But even by the early 1900s a canoe would no longer cut it. Fishing had become a commercial enterprise and to earn a living you eventually had to have a seine boat.

For most of the Native fishermen in Alert Bay that meant going to Norman Corker, the manager of the ABC cannery there. "Norm is one of my heroes," Don Pepper said. "Such a straight shooter. He dealt with the totality of the fishing experience. He knew every fisherman, every family, what was going on. He had these boats that were absolute junk and he made them fishable entities, made them available, and everybody made money—not much, but some. And so everybody was somebody in Alert Bay."

Like David Huson, a lot of these people worked at other jobs, too. David Rahn, then editor of *Fisherman Life* magazine, said, "In the 1960s a fisherman was basically a logger who fished during fire season."

When the salmon season was over, the nets were hauled into the net loft and hung there for the winter. The loft was the vast open "attic" above the cannery. At its gable end were big double doors like a barn's hayloft. Hauled up through these open doors and hung from the roof trusses, the nets had the air circulation they needed to keep them from rotting. It was here in the net loft that the nets themselves were created and mended, and ropes were spliced for the purse rings.

And it was here, in this fisherman's equivalent of a quilting bee, where the net needles flashed back and forth creating their intricate knots, that the speculation over the coming season ran highest and the conversation became most animated. The biggest sets of the season just past were rehashed in minute detail and the prospects for the coming one discussed. Would it be another lean year? Would a threatened strike see them all flat broke catching "food fish" and canning them on the galley stove?

That they would all survive somehow was not in doubt. There was firewood for the chopping; there were fish to eat; and Dong Chong, finely attuned to the local economy, would extend credit when necessary. And there were those jobs "in the woods." There was something going on everywhere on the coast. Around every corner, in every bay, there was a gyppo outfit working—a little camp on floats, a spar tree and a bulldozer crawling around yarding logs. Over the months outfits like these would get a couple of sections of logs in the water, and they usually had a job for someone who was willing to work. You didn't need a high-school diploma, just a strong back and the ability to keep your wits about you so you didn't get hurt or, worse still, killed. Dealing with the Workmen's Compensation Board took the boss away from more productive activities, and as a consequence he didn't favour the accident-prone.

But by the 1970s those catch-as-catch-can days had gone the way of the dugout canoe. The little camps had disappeared; the big ones had torn the spawning streams to pieces and used employment agencies in the city. Still the salmon had somehow survived; they were returning in larger numbers and the prices for them were going up by leaps and bounds. "It energized the fishing industry," David Rahn said. "People were getting a decent return on their capital investment and it gave them the sense that they were in business, not just scraping by. Then the 200-mile limit came in. Fishermen were directly involved in the negotiations. They were flying to Moscow, flying to New York."

The beat-up old boats and the part-time fishermen began to fall behind. To have a part in this prosperity meant having a boat bristling with technology, and to upgrade meant heavy-duty financing. The banks find neither an old fish boat nor a fishing licence an appealing type of collateral. As a consequence it was the fishing companies that supplied the financing needed to buy these big new boats. The financing came at a price, however; the companies would lend up to 49 per cent of the value of the boat, but as holders of the remaining 51 per cent they had control over the fisherman. That control prevented the fisherman from selling the boat and compelled him to sell his catch at the company's price.

"If you're a Slav," Don Pepper said, "you'll be working all winter with a company to buy a boat on acceptable terms. You can deal with banks because you've got some equity in your home and you know how it's all done."

But for most of the Alert Bay fishermen it was different. "Banks didn't want much to do with anybody in Alert Bay," David said.

There were those—Jimmy Sewid, "Hutch" Hunt and Steve Beans among them—who did buy big new boats. "They were the high-liners," Rod Hourston said. "They had another sense that the average Joe didn't have. And they were prepared to work very hard. They worked their guts out."

"The rest," Hourston said, "didn't care if they made a bundle or not. They couldn't see the sense in it. Why should they? Their lifestyle is different."

So they were left with their "crappy" old boats or remained the skipper and crew on a company-owned boat.

David said, "People think fishermen make so much money, but it's the boat owner that makes the money. If you had a company boat and you made some money, you didn't really get it; it was just on paper. We had a joke. We called it BC Packers' money."

The Yugoslavs did have their own boats. In the herring fishery they "pooled" with local fishermen—Native men who were highliners and knew where to find the fish. "These guys were in their own areas," David said. "They caught the most fish. And the Slavs packed it to town. So the locals were happy-go-lucky guys, eh. If you get somebody who lives in Vancouver to pack your fish down and you can go home for the weekend, that's fine with you."

David Huson bought his first boat, the Miss Donna, *with help from Bob Bingham.* David Huson photo

As a result, every year one of the business-oriented Yugoslavs bought another boat while the others were still tied to boat shares. David was one of the latter. Then in 1973, back in Quatsino towing logs, he got a call from Bob Bingham.

"I found you a boat," he said. "Are you still interested?" Is a fish interested in water?

The *Miss Donna* cost $126,000. She was 16.5 metres long. "For wharfage she was 48 feet. For a licence it was 57." David laughed. "They grow and shrink," he said. David put up $10,000 and Bob Bingham co-signed for the remainder. By the time he was 34, David had paid off his loan and the boat was his. He was where he had always wanted to be.

"I was born to fish," he said. He had never wanted to do anything else. But it was getting harder all the time. The dugouts and the crappy old seiners were a thing of the past. A fisherman's way of life was now a complex business requiring an astonishing array of talents. Boats could cost in the hundreds of thousands of dollars and their licences upped that price by another hundred or more. Their nets cost perhaps

$30,000. Boat owners needed a knowledge of accounting, contracts and insurance, as much mechanical and electronic skill as they could muster, a small-craft master's certificate, a grasp of any number of fishing regulations, a personnel manager's ability to attract good employees and keep them—and lastly the ability to catch fish!

The crew of the *May S.* sat around the galley table, finishing their breakfast.

"Fish and potatoes for breakfast," said Barrie. "Fish and potatoes for supper."

"And peanut-butter sandwiches for lunch," said David, reaching for a toothpick. "I was just thinkin' I'd like to go up to Kingcome village, but we'd have to anchor out and go up the river in the skiff. And I don't like leavin' the boat with nobody on board. What say we go out in the straits and see if Sonny's there?"

The winch ground the anchor up out of the depths and they retraced their route of the day before. Across a vast landscape of islands thick with trees and water like a mirror, the *May S.* was the only thing that moved. Then they were out of the scattering of islands and into the broad reach of Blackfish Sound. It was the sunshine that made them pause. Only a local can appreciate the rarity of a week of September sunshine up around Johnstone Strait. The *May S.* sliced through the water then slowed and then stopped, and all the small sounds destroyed by the thudding of the engine were once more audible. There was the soft slap of water against the hull and the squealing of seagulls wheeling over a tide rip somewhere. The two men came down from the dodger and settled themselves on the bulwarks in the sunshine.

"These last few years you really had to hustle," David said. "I used to figure on a 24-hour opening I'd be up for 42 to 44 hours. We'd go out there and get to my spot and we'd start fishing at 6 a.m. on Monday. We'd have to fish right through to 6 a.m. on Tuesday. You had to fish all night to compete. Then you'd have to run in and sell your fish and pick up fuel and water and some grub. And then you'd run back out to your spot to anchor. By this time you'd be third or fourth in line for the best spot."

But anchored there waiting for the next Monday's opening, life was pretty good if you had the right crew. "If you had somebody on board who wants to party it was miserable for everybody," David said. "That's why it's gone to families. But with the right crew it was great. You have outboards and fishing rods ..." For after seining for 24 hours straight, what brought the crew the most pleasure on their days off was catching a fish, of course. It's called salmon fever.

Sometimes fishing farther north seemed a better bet. One year David, who was living in Victoria by now, was getting ready for the season to open at Prince Rupert when the DoF moved the opening date up two weeks. His crew tumbled aboard and for a fee of $75 some nameless operator lifted the span of the Johnson Street Bridge into the night sky. The *Miss Donna* slipped through and rushed past the bluffs of Beacon Hill Park.

"We were in a hurry. We were trying to make it but we never did make it. We got as far as Bella Bella and I was just beat. I had a new crew, nobody on there that really knew that part of the coast, and I just didn't want to push it in case we piled the boat up."

David's new crew gradually became his trusted crew. "I had lots of good guys with me and they stayed with me," he said. "I worked them pretty hard because I figured the net wasn't making any money sitting on the drum. You have to do that now with the short openings. But these guys were good. They were always behind me. They knew I was trying to pay the boat off."

"And now," David said, "they're talking about the salmon disappearin' altogether. This last summer they had one opening and no fish—I mean 300 fish was a big deal."

The tide was turning. David lined up a spot on the far shore to calculate their drift.

"My nephew's out diggin' clams. You even have to have a licence for that these days. They go out in his little gillnetter. Him and three other young guys. They chip in for the fuel. Their grandparents are always worryin', always trying to find out where they are. They've got a cell phone on the boat but they never use it." He laughed.

Barrie was looking over the side, down into the depths, wondering if there was a cod down there somewhere. "Can they make a living at that?"

"I wouldn't call it a livin'. They're happy they made a few hundred dollars. They don't have a chance to do anything else. In Alert Bay they told me the young guys got a job for the government cutting salmonberry bushes. Next year the same thing. They said, 'Never mind giving us a job if that's all we're going to do.' Then the government got the idea of painting the houses—fancy new colours. If you bought the paint they'd pay the kids to paint your house, but you had to use these colours. They brought a guy up from town to pick all the colours. He was a colour consultant. You wonder what one of these guys makes. Anyway they were goin' to make it look like Petticoat Junction. What are they tryin' to do? It's awful to talk to people; everybody in the place, their savings are all gone. You get depressed."

Five-day openings were now part of the dim distant past, and the 24-hour openings that Rod Hourston declared a minimum had shrunk once again. And packing fish didn't cover its costs.

In the old days when the fishing grounds up north and down south were closed, everybody went to Johnstone Strait. Perhaps 100 seine boats and 300 gillnetters dotted the water there and "in back," and on those beautiful August days everyone was making money.

"Now they've got it down to 15 hours," David said. "I don't know what's going to happen."

"What's going to happen," said Barrie, "is one boat is going to catch all the fish in Johnstone Strait and send it off to Japan for sushi and that's that."

"I let all my crew go in June," David said, "'cause we couldn't pack any fish this year and it wasn't fair to them. That's it, you know; it's not just yourself, it's five other guys you're makin' a livin' for. A coupla my old crew bought gillnetters, so they might be okay. And one of my guys is tryin' to make a living divin' for sea urchins in the winter off the west coast of Haida Gwaii."

"That's kind of a brutal job."

"Yeah, and he's a really good guy and he just got married awhile back. They got a kid now so I feel pretty bad about it but what are you supposed to do? But you know what it's like in Haida Gwaii in November. It gets pretty shitty out there. These guys fish out of aluminum skiffs. They use dive tanks. They got a bag made of webbing and they go down for 20 minutes each dive and rake the urchins into the bags and hand them up to the guy on the tender boat and then go down again."

"It blows 100 miles an hour up there in the winter," said Barrie.

"I know it but what are these guys supposed to do—flip hamburgers? They're their own boss. They got their pride."

"What do they do with the urchins, then? Sell them in Rupert?"

"They use that seiner *Ryoi* as a packer," David said. "You remember her? She's the one that was used as a mother ship selling whisky to the Americans in Prohibition. She's still in nice shape. The guy that has her now put all these freezers on board, so when she's loaded they take her to town and ship the stuff to Japan. Supposed to be a delicacy there.

"So then the guys leave and you go out and you've got a green crew. And that's when you have accidents."

Barrie was jigging a line over the side. Suddenly he began to yank it in. "Big son of a gun," he grunted.

David leaned over and looked down into the water. "Nice big ling," he said. "I like ling. I'll get the knife out of that box and clean it."

As the fisherman, Barrie was prepared to clean his own fish, but David was firm.

"No you won't," he said. "I seen how you used to fillet fish and I doubt if you've improved with age. You were a butcher—left all the bones in."

Drifting there they might have been the only living beings in this world. They were not, however; they were being closely watched. Their every movement since they caught the fish had been noted by a pair of sharp eyes. The instant David tossed the entrails overboard there was the woof, woof, woof of metre-wide wings and the eagle swooped, snatched and was gone.

The coast is still a wild place, Barrie thought.

CHAPTER 9

High Boats:
Barclay Sound — Fred Joliffe (BCP)
Ermelina — Louis Benedict (ABC)
Qualicum L. — George Cook (Can Fish)

Pioneer Journal, August 26, 1953

There are some 2,600 salmon streams in the province of BC. They trickle down from the mountains, tiny rivulets twisting and turning among the trees, gathering volume as they go. At a natural depression they accumulate and become a lake, and then, at some point, they burst the bounds of the lakeshore and rush on downward to the sea. They are creeks now, but their presence is still hidden by the canopy of the forest. Their exquisite flow of water, gurgling around mossy rocks and ancient trees hung with lichen, is a secret until they broaden out into rivers and separate the forests that line their banks. At least that was the way it was in the beginning.

The Nimpkish is one of the most important of these rivers. Each spring fingerlings emerge from eggs buried in the gravel bottoms of its streams and pursue their separate life cycles. The coho and the springs spend the second year of their lives in those same streams, and the sockeye find their way to Nimpkish Lake and live out their second year there. Then all three species head downstream and into the open ocean. "Off they go for two years," said Rod Hourston, "or three years in the case of the springs. And we have no idea what their survival rate in the ocean will be. The first feel we get for the size of the run is when the nets hit the water."

Thus the 1957 spraying program presented no threat to the 1958 run, which was safely out at sea at the time. When it did return the fish came in record numbers. Fishermen still speak of 1958 in glowing terms. The sunlit waters of Johnstone Strait and Blackfish Sound were dotted with seine boats that summer. They were there, tiny coloured dots along the shore where the Vancouver Island mountains dropped into the sea. They were there on the glassy waters of Blackfish Sound, off Pulteney Point in Broughton Strait, in the maze of little islets that formed the Merry-Go-Round.

No one can tell you exactly how many boats were in this fishing fleet. As Steve Warren said, it might start with 40 boats. When news of the big run leaked out of the radio telephones that number could double or even triple.

Rod Hourston said, "It's tough to estimate the number of fishermen that end up in Johnstone Strait because it's a mixed fishery that involves the Fraser River fish and all the salmon from the inlets. One thing for sure, once the sockeye hit Johnstone Strait, all the fishermen want is those red fish. They all head down—they leave the pinks—and head down to fish sockeye."

"Someone comes on the phone," remarked Billy Proctor, who has fished these waters all his life, "and says 'Sockeye is going to open!' and all hell breaks loose."

It was the DoF's job to manage the fishery—to measure this season's run against those of other seasons and decide on the days each week that it would be open to fishing. It was, of necessity, an inexact science. The department had accurate records of past catches but the number of sockeye arriving each year was an unknown until the first catches were tabulated. So the boats were required to give the fisheries officers their numbers at the end of the day. In the operations room the catch per day of every year by district was plotted. By comparing day one of the current year with the same day of previous years it was possible to get a feel for the size of the run.

The other variable in this management was the size and efficiency of the fishing fleet. As the gear improved and the technology advanced, the catch per boat increased. There were different types of fishing to be considered, too. Gillnetters, trollers, purse seiners, gillnet trollers and the Native fishery all demanded a portion of this resource.

The blunt instrument that regulated all this was time, and deciding how much time to allow the fishermen was almost a day-to-day activity. "How big is the run? What are they taking? How much are they taking? When do we close it? How's it looking? Could we give them an extra day?" When fishery officials huddled, these were the questions that their consensus of opinion attempted to answer. "It was so critical," Rod Hourston said. "The fleet was so big and the gear had become so efficient that if you gave them an extra day you could really hammer the resource."

And as if this pressure on the fishery was not enough, out in the Pacific there were foreign fishermen intercepting the salmon before they even reached the BC coast. "For a long time we only had a three-mile limit, which is a piece of cake for those guys out there," said Hourston. "And then we got 12, which is still nothing. Finally we got the 200-mile limit, so now our groundfish and herring stocks are protected because the shelf doesn't go out that far."

In 1946 when Rod Hourston was a student working at a summer job on the Skeena, it was a five-day fishery. It was the same in the 1940s when Steve Warren and David Huson's dad, Spencer, bought their own boats, and in 1957 when Barrie and David stepped off the *Cardena* and became fishermen. But by the time Hourston became director of the Pacific Region in 1960, the efficiency of the fishing fleet had increased to the extent that the openings had dropped to two days. "We'd usually open it for two days and take a look," said Hourston. "Sometimes we could give an extension. But sometimes we'd only open it for a day. That was the minimum. You can't have anything less than a day in a fishery."

The fishermen of Alert Bay felt, for the first time, an intimation of trouble, a flicker of anxiety in their guts. Perhaps salmon were not, after all, an inexhaustible resource that would always be there in the waters around Cormorant Island. "It was a gradual thing," said George Cook. "You didn't notice it for awhile but that's what a few people were thinking."

To make a living, of course, a fisherman must fish. With the openings for the salmon fishery ever shorter, and the herring fishery closed, they turned their attention to other species. Following the Norwegians' example they fished for halibut. "Halibut fishing is a hard, hard job," Sam Hunt said. "Right out in the big swell. It's a matter of how much gear you can handle and where to put it, because it's spotty—the fish aren't everywhere. And the boat's rolling all day long, a heavy roll. You've got to be born to it.

"After the halibut we trawled for dogfish. We weren't professional trawlers but we loaded the boat. We fished four or five miles off the Charlottes with the Americans. The competition was rough. A trawl is 125 feet long and 80 feet wide. It drags on the bottom and what it doesn't catch it crushes. To my way of thinking it's the most destructive thing on the coast. We'd be out in Hecate Strait—10 boats side by side. It catches everything; the deck of the boat was solid baby halibut six inches long."

Twenty-five years later the fishery continues. It isn't the voice of a single Native fisherman protesting its destructiveness now. It's the Sierra Club. "There are still no catch limits for most species being caught by trawlers on Canada's west coast," says the Sierra Club of BC website. "The fisheries minister is failing in his duty to conserve the fisheries resource … everywhere we looked the minister was approving fisheries despite scientific evidence of ongoing species decline."

In 1968 the fisheries minister of the day, Jack Davis, was too busy to concern himself with trawlers, for he was introducing the concept of licence limitation to the fishing industry. This was an idea that had been around since 1961. Because of the increased gear efficiency and the declining stocks, the department had managed the industry with the only tool it had: fishing time. Now it decided to add another tool. The size of the fishing fleet would be cut back to the point where the remaining fishermen could earn a decent living. Commercial fishing was to be, in effect, an exclusive club, entry to which required an A or B licence. An A licence was issued to those fishermen who had landings over 4,500 kilograms prior to 1968; B licences were for those who had landed less than that amount. The licences went to the boat, not to the individual. The B licences would automatically expire in 10 years. Those forced out of the industry because they couldn't meet the aforementioned requirements would have their boats bought back by the government. Canada was one of the first countries in the world to try this approach, and somewhere in the theory the practical side of things got lost. As a result the program created myriad problems, including the unintentional propagation of millionaires.

To begin with, a program that was meant to "limit" started off by enlarging instead. "As soon as the licence limitation was announced," said Rod Hourston, "the halibut boys got into the act. At that time we still had a big halibut fishery up in the Bering Sea. These guys said, 'We can't fish salmon but the salmon fishermen can fish halibut. So why shouldn't we get an A licence?'" Much against Hourston's better judgment the government caved in, and this opened up another big fleet to the salmon fishery. The salmon fishermen, in turn, demanded halibut licences.

So the fleet was still too big. The Pacific Salmon Commission calculated that 100 to 150 boats was the optimum number for the coast fisheries. There were, and still are, 300 in the south area alone. "Such efficient fishing machines," Ian Todd said. "And faster. Before they picked an area and fished it, but now they can race around to wherever they think the fishing's good."

"So then we had to limit the size of the boats," said Hourston. It was at about this juncture that an old fisherman accosted him one day. "You're never goin' to beat us," he said, "because we have all the time in the world to beat you. We're sittin' out there night after night figuring how we're goin' to beat your regulations." True enough, thought Rod Hourston, ruefully.

The trouble was that fisheries regulations were formulated by bureaucrats in Ottawa who had no familiarity at all with fish or fishing, and they issued edicts that were full of loopholes and often patently ridiculous. With no confidence that this bureaucracy knew what it was talking about or had any real concern for the fish themselves, fishing became a question of who could outwit whom.

The next step in licence limitation was the "buy-back" program. Fishermen retiring or leaving the industry for any reason could sell their boats and their gear to the government, which resold them on the understanding that they would be taken out of service as fish boats. Nothing, however, prevented US fishermen from buying good boats from the buy-back program and using them in US waters, where they often intercepted salmon headed for BC. What's more, the funding for the program was inadequate. It had been assumed that the licence fees paid by the fishermen would fund the buy-back. But the law of supply and demand, something else the bureaucrats had overlooked, had taken over. "In the late 1960s a not-too-bad, 30-year-old wooden seiner with a net and an A licence sold for about $40,000 to $60,000," Alert Bay fisherman Mike Weigold said. "From 1973 to '77 it was closer to $100,000 to $140,000. By 1981 they were up to $150,000 to $200,000. By 1990 the licences became the only real value; that was one of the largest boat-replacement periods and by then the licences were worth $400,000 to $500,000."

With a licence worth hundreds of thousands of dollars, fishermen were replacing their junky old boats with efficient new ones that were capable of catching many more fish. The artificial manipulation of market forces resulted in unexpected and undesirable effects; it confused the innocent, provided windfalls for the shrewd, created appeal boards and much bureaucracy—and cost the taxpayer.

Halibut licences and those for cod, sea urchin and herring were also issued with the proviso that the possessor had to produce a minimum catch in order to retain them. In the case of the herring fishery there were even more complications. In the herring fishery the licence went to the individual, not the boat. The companies sought to get as many of these licences as possible and then use them as bargaining chips. They would transfer one of their licences to a fisherman (or supply a second one to a man who already owned one) at a negotiated price that was often less than the going rate. There is, of course, no free lunch. In return the fisherman was obliged to sell his herring to the company—regardless of the price they offered. And sometimes the price offered was less than that of competing companies. Occasionally fishermen actually ended up taking a loss on this arrangement.

The losers in this forest of regulation were the Native fishermen. With their "crappy old boats" few of them could catch the fish that the newer boats did, so fewer of them could meet the requirements for licences. "So all of Alert Bay lost their licences, eh," David Huson said, "except one guy and he had a fire on his boat and had to sell his licence to pay for repairs. Now no one in Alert Bay has a halibut licence. I had seven different licences on the *North Isle* and they said, 'you're not using them' and took them back. And now they've all turned valuable. A cod licence is worth $80,000 to $90,000 now." It was all so complicated and hedged about with regulations—incomprehensible to those who had started fishing when reputation and hard work was enough. "When I first went to a fishing company with my dad it was just a handshake deal," said David.

And then, in the middle of all this licence limitation, word got round of a new fishery—something that promised unbelievable profits for fishermen. "Some northern outfit was selling herring for the roe," said Hourston. "It came from out of the blue. We had never heard of it before. Having almost destroyed the resource we were very jumpy about this new fishery, because the difficult part is the fishery takes place just as the fish are about to spawn. It's a critical time. So we were pretty careful right from the start. But it became an incredible fishery."

Regional fisheries officers protected this new fishery with a strict quota system. Once the harvesting had reached a pre-arranged quota the fishery was closed—sometimes less than 24 hours after it opened. Trying to defend the supply of fish against the demands of an ever larger, ever more efficient fishing fleet, "all you can do is crank down on the fishing time," said Rod Hourston.

Such a tightly managed fishery is more like a fishing derby than anything else. When the DoF fired the starting gun it unleashed a competitive frenzy. By 1979 (the year the department metamorphosed into the Department of Fisheries and Oceans, or DFO) herring prices had topped $5,500 a tonne. The boats and their skippers, spurred on by stories of catches worth $1 million, bore down on the fishing grounds like the tanks of an advancing army. Guided by their sonars, they jockeyed for position and fought for space to make a set. And then, in

CAPE JAMES

PREVIOUS PAGE AND ABOVE: *Fish boats became bigger, faster and more numerous and their technology more sophisticated. It all added up to a devastating assault on the resource.* Photos courtesy UBC Library Rare Books & Special Collections, Fisherman Publishing Society

the few hours that the fishery was open, they drummed their huge nets in, lowered their pumps into the mass of fish and sucked up their fortunes—enough fish to leave the decks awash and sink them if they were caught in bad weather. And there was plenty of bad weather. Many of these overloaded vessels sank, taking their crews with them. And the herring—those little silver fish that were the promise of unbelievable riches—went back to the sea they came from.

"When the roe herrin' came in everything changed," Steve Warren said. "Exceptional big seasons, moneywise. Of course everybody didn't make a million; there was the rich and the poor—same as always."

Steve was 65 the day he went out on the roe fishery for the first time. He went to Barkley Sound with Gordon Wasden on the Cape Canso. They were at sea exactly one month and Steve's crew share was $21,000. "Most I ever made in that length of time," he said, after a lifetime of fishing.

Ian Todd had left the DFO by this time and was working for a relatively small fish company. "It was crazy," he said. "A sort of love-hate relationship. My briefcase habitually had a quarter of a million dollars in it—and we were small potatoes as far as fish buying went."

"It just became meaningless, in a sense," he said. "I remember standing on the pontoon of a Cessna out in Georgia Strait. It was so windy we couldn't get alongside the boat that needed the money, so I just grabbed a canvas bag with $50,000 in it, wound up and flung it at the guys on the boat. It wasn't 'til it was flying through the air that I thought, 'Wait a minute …'"

Now fishermen with pocketfuls of money went to their accountants and were advised to create tax write-offs—and the best write-off is a new boat. The tax regulations provided every encouragement, the write-offs were incredible, and as a result, money from the roe fishery expanded the fleet from 250 to 550 boats—and that, in turn, was the fleet that fished for salmon in the summer.

Money from the roe fishery helped David Huson. Like Steve Warren, David remembers leaving one March 1 and returning March 23 with $29,600. But both the *May S.*, his father's boat, and the *Miss Donna* were small for this fishery; they packed 35 tonnes compared to the bigger boats that could carry 100. With money like this coming in, David thought, I could buy something bigger. He owned the *Miss Donna* free and clear; now he mortgaged her and contracted with BC Packers to buy another boat. The *North Isle*, with her herring and trawl licences, cost him $597,000. He had good crews on both boats—good fishermen who routinely loaded the boats to the gunwales with herring.

An overloaded herring boat, of course, is further burdened with a pile of net on her afterdeck, and on top of that a big herring skiff. One day off Cadboro Bay this became too much for the *Miss Donna*.

"Reg Glassco was running her," David said. "He had an experienced crew from Bella Bella—Eric Wilson on the wheel. And suddenly Eric noticed something funny about the boat. It was going over. Lucky he's a very calm guy. He kept going in a straight line, running half speed. I was following behind and the first thing I noticed was the skipper out there on the back deck wading around trying to get the net off. The stern was going under but they got the net off and it started to come up. Those guys saved the boat."

There was a worse disaster to come, however, one that seamanship could not prevent. "The first year we went on strike and I couldn't go

David Huson mortgaged the Miss Donna *and bought the* North Isle *(above). But the double whammy of a fishing strike and usurious interest rates cost him both boats.* Photo courtesy UBC Library Rare Books & Special Collections, Fisherman Publishing Society

fishing," David said. "So that was tough with a big debt like that. And then the interest rates went up to 24 per cent. I thought it out. I've got a family in Victoria. I've got a house there. I need my house."

It was that simple. He lost both boats.

There in Blackfish Sound the *May S.* was a small white speck on a stretch of brilliant blue water.

"This sure beats fishing herring in January," Barrie said. He dropped a bucket over the side, hauled it in and heaved it over the deck. The last of the ling's blood sloshed out through the scuppers.

"You bet," said David. "All the lines froze hard and stuck to the deck, so you have to kick 'em off before you can do anythin'. And then

one time I got a crack in the hydraulic fitting and the drum wouldn't budge. The fishery guys were there and you only get one shot at it, so that kind of left us buggered up.

"I remember Silas on the *Pearl C.* In all my years I never seen him put a jacket on. Didn't matter if it was snowing, he'd sit on the bench behind the dodger for 16 hours a day and he never once said anythin' about bein' cold. Those old guys from the Bay were pretty tough."

On the north coast sunshine is not to be confused with heat, of course. The frigid ocean sucks the heat right out of it. Still, like Silas, you get used to it.

"Can't go to town in weather like this," said David, wrestling around behind the winch. He extricated two aluminum garden chairs, unfolded them and put them out on the freshly washed deck.

"First class all the way," said Barrie.

"Yep. This always was a family boat. We done a lot of cruisin' around in her." David lowered himself into the tattered webbing. "In the spring I'd take the *May S.* up to Maple Bay and get her pulled out. We'd put a coupla cars on the back deck before we left. My dad's little Toyota truck. And about 20 people would get on board and we'd just go cruisin' around. Go up to Maple Bay, drop the cars off and go up and buy an ice cream cone. Then we'd drive back down to Victoria and they'd clean her up and paint her." He butted his cigarette in the empty salmon tin balanced on his knee.

"When I was a little kid," he said, "I thought the *Lake Como IV* was a real big boat. But that's what a kid thinks; she was maybe 45 feet long. So about 1960 my dad started looking around for something bigger. There was the *May S.* for sale and a steel boat called the *Tibur*. They wanted $40,000 for the *Tibur* and my dad said, 'Forty thousand—we'd never pay it off.' So they gave him a good price on the *May S.* and he bought her.

"I grew up on the *May*. I was 18 when we got her. We fished halibut. And when we weren't fishin' we got towing jobs and things like that. We worked for Laurie Creelman a lot. We were workin' for him up in Winter Harbour when they had that earthquake in Alaska in 1964. A big tidal wave come in there. It just wrecked everything. You wouldn't believe it. The current was runnin' full swing. Just boilin' everywhere. If you're ever in Winter Harbour—you know, the town of Winter Harbour—in a tidal wave, go up to the end dock, the one right up at the far end. Don't go to the BC Packers' one."

Barrie had pieces of the oil stove spread out on a newspaper on the hatch cover. He picked a bit of sludge from the carburetor. He thought it unlikely that he would ever find himself in Winter Harbour in a tidal wave. Still, sitting there swapping stories in the sun was as happy as he'd been in a long time.

"You know, you couldn't buy a day like we just had," said Barrie. "Have you ever thought of doing this kind of thing when you retire? You could extend the house out over the hatch and fit her out for those ecology trips. If you took people out and showed them an eagle and a few whales and they caught some fish, you could probably charge a couple of grand a week."

David wasn't listening—or perhaps he preferred not to. After awhile he said, "I'm a fisherman, Barrie."

CHAPTER 10

High Boats:

Chief Takush — Simon Beans (BCP)
Sea Angel — Jimmie Wadhams (ABC)

PIONEER JOURNAL, SEPTEMBER 13, 1950

The *May S.* proceeded briskly past the end of Hanson Island and into Blackney Pass. On her port side, just at the tip of Cracroft Island, there's a tiny bay where a half-submerged stone wall, the remains of an ancient fish trap, are still faintly visible at low tide. The fish came in at high tide, spilled over the wall and were trapped when the tide receded. Thousands of years ago fishing places like this were jealously guarded family assets, handed down from one generation to the next. The owners of this particular site, the Prevost family, were fortunate; the tides that race through Blackney Pass carry fish food—and fish—with them.

In 1924, generations later, this area was still a prime location. At the north end of Harbledown Island, the island that borders the east side of Baronet Pass, there's a lagoon. Here the Cook family had a fish-buying camp and Stephen Cook's capable married daughter was in charge. For two and a half months each summer she and her six small children lived there in a wooden shack, part of which was partitioned off to serve as a little grocery store. There was a float in front where the dugout canoes came in with their catches and the powered boats bought gas. The bounty of Blackney Pass could earn a man and wife, trolling from their dugout canoe, enough in a season to see them through the rest of the year. Their catches were picked up by a packer and delivered to the Canadian Fishing Co. cannery at Bones Bay. The

21913

six small Warrens, one of whom was Steve, served in the store, pumped gas, fished from their little skiff and had the happiest time of their lives.

But Steve's mother died in 1927 and 75 years of incessant rain had reduced the shack to a few pulpy bits of wood lost in the underbrush. So when the *May S.* slipped by on that bright September day nothing remained of the fish camp but the lagoon itself.

On a strong ebb tide the *May S.* raced through the pass dodging bits of driftwood and islands of matted kelp. A seagull, webbed feet planted firmly on a chunk of log, whirled gravely along this moving highway. Farther out in the middle of Johnstone Strait 11 storeys of cruise ship rushed northward. Those on the *May S.* peered through their battered binoculars.

"Nobody at all on deck," said David.

"All inside attending a lecture on the flora and fauna of the BC coast, I guess."

David's interest in the cruise ship had waned. "You know about the blackfish and stuff, livin' up here and workin' with Steve and Eddie and all that," he said. "Blackfish've got a whole different meaning for us—always have, you know. But now it's like a circus up here sometimes. They've got those kayak expeditions. The people pay to camp on the beach for a week. Shit, they're travellin' across the straits in the fog in August and you can't see them on radar." He was staring through the binoculars. There in the sunshine near Blinkhorn an aluminum drum seiner lay beside the circle of her purse seine.

"*Western Moon*," said David. "Doin' a test set. We'll go over and say hello." For a moment, as they got closer, it was just as it used to be; the drum was revolving bringing the net in with it—and just for a few moments the crew were alert and happy, peering down into the depths trying to judge the size of the catch. It wasn't as it used to be, of course. This was a "test set," a single set carried out in a specific area

PREVIOUS PAGE: *The drum seiner* Western Moon *conducts a test set in Johnstone Strait.* From the collection of Barrie McClung

for the DFO. It was made in order to estimate the number of salmon in Johnstone Strait. In 1958 a boat down near Adams River caught 6,000 fish in one set. One of her crewmen, a boy of 16, bought himself a brand new Pontiac with the proceeds. Now the watchers on the *May S.* saw the *Western Moon* pull 60 fish up out of her seine net and the fisheries inspector gave one of them to David and Barrie.

The DoF began doing test sets around 1962; the sets were, and still are, used as an index of abundance. To carry out these first sets the department chartered an Alert Bay boat belonging to Jimmy Sewid. This choice was deliberate, for Jimmy was a wise and respected member of the Native community. Instead of his regular crew he brought along four or five First Nations skippers to crew his boat, and he explained the reason for this to the biologists.

"I want them to see what's going on," he said, "because they don't really trust you people. All they see is restrictions. They think you're out to do us in. You'll have to explain what you're trying to do."

That was the beginning of some productive—and heated—discussions. Ian Todd was still with the DoF then. "We kept the lines of communication open," he said. "We had public consultations and some big arguments but all in all we had a pretty good relationship. I remember Chris Cook standing up at a meeting in Campbell River one year. 'How many fish would you have to catch in a test set to open the fishery?' I said, 'Five thousand.' I pulled that figure out of thin air actually, but I had to give them the impression that we had a long way to go. I wanted them to realize that we had to get fish on the spawning grounds or they could forget it. And there were years when the test boat *did* get sets that big.

"Fishermen are great conservationists from December 15 to June 15," Todd said, "and that doesn't bother me. We'll work together in the winter to come up with a plan. But once the whistle blows it's your job to get as many fish for your company as you can. And it's my job to make sure you don't get out of hand, because my job is to get enough fish through to the spawning grounds ... so that we can have this glorious fight all over again two, three, five years hence."

But by the 1980s and '90s this spirit of co-operation was in tatters. "The local fisheries officers were good people," said Todd, "but they had so much heat put on them. Their job has gotten so complex. Fly over the mouth of the Fraser now and you'll see no vacant foreshore; it's all docks and industry and housing developments. So the jurisdictions are split. They're trying to deal with municipal, regional and provincial districts and a whole lot of different agencies. The complexity of the DFO's job has increased *enormously* since the late '70s."

The complexity had increased and the leadership, the fishermen felt, had deteriorated. When Don Pepper left the fishery in 1969 he donned his economist's hat and went to work for the Department of Fisheries in Ottawa. What he found there appalled him. "The department didn't know fish. They would vaguely talk about it but they didn't really have a clue."

They were not averse to having someone around who did, however. The minister of the day, Roméo LeBlanc, discovered that the economist he had hired was "a guy who fished," and in short order Don Pepper had solved some of his thornier problems.

"So I was Roméo's fair-haired boy," Pepper said. He knew fish; now he would learn something about government. "I would fly out to BC with Roméo on the Jetstar and I'd be all excited about what they could do out here."

And Roméo would say, "Calm down. There are no Liberal seats out here."

But it's never a mistake to please the public, as any politician will tell you. A series of informal hearings held around the province in 1977 had made it clear that the public in BC overwhelmingly endorsed the idea of salmon enhancement. "At that time everyone in fisheries who cared about fish saw salmon enhancement as the salvation," said Don Pepper. "To get the money the program needed, the DFO people in Ottawa (and I was one of them) fudged the numbers. We got the money by hook or by crook, and those working on salmon enhancement became the elite."

Then things began to change. The "class of '49," those ex-servicemen who had joined the DoF after World War II, had worked their way up the organizational ladder and knew "everything and everybody," retired. "In the Pacific Region they lost the leadership and they lost the vision. By the time Rod Hourston retired that was the end," said Don Pepper.

They were replaced by whiz kids from management schools. "Modernization" was their buzz word and computers were their gurus. Seduced by ever more sophisticated computer programs, they substituted an intellectual and technological exercise for knowledge and issued edicts that were sometimes so ridiculous that the regional director at the time, Glen Geen, refused to implement them. "He was fired," Don Pepper said, "and things went downhill from there.

"The problems were too tough for them," Pepper said. "They didn't understand the licence limitation. They didn't understand the buy-back program. They started that program with $2 million—$2 million! That's two boats! They hadn't a clue. Nobody understood the fundamental issues. Nobody had the broad view. More important, no one was accountable.

"This is a huge bureaucracy whose leaders are never seen because, like Gilbert and Sullivan's Duke of Plaza-Toro, they are leading from behind. Today's fishermen are fighting shadows."

Mike Weigold is, or rather was, one of the fishermen who battled these faceless bureaucrats. His father emigrated from Germany in 1952 and came to Alert Bay shortly afterward to work as a diesel mechanic in the shipyard there. Born in Alert Bay, Mike had his first trip on a seiner at seven. At 12 he had his first paying job working for two weeks on the *Izumi II* washing dishes and "plunging." By 15 he was a full-share crewman. He worked, over the years, as skiffman, cook, engineer, beach man—all the jobs there were on a seiner (although in the beginning his skipper gave Mike half his wages and gave the other half to his father "so it wouldn't get spent").

By the spring of 1977 Mike had his own boat, the *Bruce Luck*. Full of piss and vinegar and the happy belief that anybody can do anything, Mike has a quick infectious laugh and a proclivity for daunting projects. In 1985 he discovered that the *Bruce Luck* had what he blithely describes as "a little rot problem." Together, he and his

The Bruce Luck *in Alert Bay in 1983. Under the broiling heat of greenhouse plastic, Mike Weigold, his wife Lee and their friends repair 'a little rot problem.'* Mike Weigold photo

wife, Lee, and his friend, David Rahn, built a floating shed over the aft half of the boat and sheathed it in greenhouse plastic. "Man, did it get hot in there when the sun was out," he said. Then they stripped off the fishing gear, drum, winches, mast, boom, stern roller and net-spooling gear. The decks quite literally cleared, they tore the bulwarks, deck and beams out from cabin to stern, replaced them and put everything back together again. Like a house renovation, this kind of project often leads straight to the divorce courts, but in this case all participants emerged on friendly terms. Mike and Lee, neither of whom look old enough to have an adult son, continued to fish together and to upgrade the *Bruce Luck*. By the 1990s she had a 275-horsepower Caterpillar engine, a pursing winch, a coloured sonar, a computer with GPS and a tank system that allowed them to unload in 20 minutes. It was possible for them to fish in 25-knot winds.

"You can do it," Lee said, "but it isn't fun."

"It's the power," said Mike, "the hydraulics—all standing rigging."

Like David Huson, Mike was "hooked on fishing" and saw it as his future, but his lack of confidence in the DFO's management made that future uncertain.

"From a fisherman's point of view you want to be dealing with someone who knows what he's talking about," said Don Pepper.

Fishermen no longer felt this was the case. Fishing regulations appeared to be increasingly arbitrary, and a political agenda seemed to have replaced any real attempt at conservation. Runs that were endangered—that the fishermen themselves didn't want to fish—were open and, conversely, healthy stocks were closed.

The fishermen from Alert Bay stopped trying to co-operate with a government agency that appeared to them to lack integrity and, instead, engaged in a game of cat and mouse. "The part of the mouse is very challenging," Mike Weigold said. "The DFO doesn't have any idea how imaginative fishermen can be." In 1975 and '76, for example, the Alert Bay fleet fished Parson Bay for springs. The DFO realized that the springs went into Parson Bay and closed it. "So we just followed the fish around the point and caught them there, where it wasn't closed," Weigold said.

As satisfying as it was to best the bureaucrats, the fishermen realized that they were, in the end, only cutting their own throats. The answer, of course, was a co-operative effort, the kind of dialogue that had once taken place between Ian Todd and Jimmy Sewid and Chris Cook.

In those days if you brought something to the attention of the biologists it was acted upon. But now the biologists were, if you'll pardon the expression, low man on the totem pole. The long-standing problem of the pinks was an example. For years the DFO had divided Johnstone Strait into blocks, which it opened and closed in an effort to preserve the pinks. To the fishermen this was ridiculous because they knew, as the DFO apparently did not, that the pinks didn't frequent the Vancouver Island side of the strait. The Alert Bay fishermen decided, among themselves, that the answer was a "ribbon boundary" for upper Johnstone Strait: a boundary that ran, not across the strait, but up and down its length. Their attempts to negotiate this change with the DFO were futile, so the fishermen put the regulation in place themselves. A group in Alert Bay talked it up on the radio, explained what they were trying to do and effectively closed the mainland area to fishing. The whole fleet stayed out of that side of the strait. A group of fishermen ("the troops") had identified a problem, found the solution and then enforced that solution—all on their own. David Rahn said, "*Years* afterward the DFO brought it in as a regulation."

In the 1930s and '40s Dave Cameron was the fisheries officer in Alert Bay. He was the proverbial dour Scot, a big heavy man who lumbered down to his little fisheries vessel and kerchunk-kerchunked around the area handing out fines and warnings for minor transgressions. He was regarded with affection by the community. In 1952 in Thompson Sound, a fisherman actually assaulted a fisheries officer and was hauled into court, but this isolated incident didn't reflect the tenor of the times. By the 1980s and '90s the antagonism between fishermen and the DFO, though it may have been less physical, was more pervasive. Fishermen saw the industry being managed by what they termed the AEBs—Arrogant Eastern Bastards—and the mood was one of confrontation. Things only got worse.

"In 1994 the fishing fleet caught 1.2 million fish in one day," Mike Weigold said. "We had become so efficient we could catch a whole run, and nobody to stop us—nobody who understood what was happening. The dynamics of the regional office had changed. It had become just a facade." Frustrated fishermen beat their fists against that facade but their complaints were "tabled" and their suggestions judged unworkable. For example, the fishermen proposed clearing Charles Creek of debris and re-establishing a remnant run. They would take salmon eggs from Viner Creek, incubate them and replant them in Charles Creek where the run had been wiped out by logging. But a couple of DFO people cited transplant implications as the reason they wouldn't give the go-ahead.

Fishermen are independent by nature and these ones were frustrated beyond endurance. They formed the Johnstone Strait Salmon Fishermen's Enhancement Society and decided to go ahead with their

The Charles Creek Project: Chum are netted in Viner Creek for their eggs, which will be transplanted to restock Charles Creek. Mike Weigold photo

proposed project—with or without the DFO's approval. It was early fall and time was of the essence. A few weeks later a flotilla of nine seine boats backed away from their moorings in Alert Bay and headed down the strait, white water peeling away from their bows. There were plenty of deckhands to coil up their lines and yank their bumpers up over their bulwarks, for the boats carried 53 Alert Bay fishermen—Native and non-Native alike. They also carried skiffs, power-saws and a planting device that Dick Johnston had made out of used parts. Dick Johnston was one of those endlessly ingenious people you meet a long way from a city. He had fished, worked in a mill and in logging camps, and built his own shipyard. Presented with a request for a complicated and expensive device he had never seen, he said, "Piece of cake."

"It worked great," Mike Weigold said. "Trouble was, we didn't know exactly how to go about it, and the fisheries threatened to fire anyone who got involved. They said you can't do habitat work without an engineering study and a permit."

"It's really easy to lump the DFO into one composite bad guy," said Mike Berry, a biologist in Alert Bay. "There were people in the department—George Bates, John Lewis and a couple of others—who were very supportive. Their attitude was, 'Screw the permits. You guys need to do it. Go do it.'"

But others didn't see it that way.

"There was a stupid conviction on the part of a few people," Mike Berry said. "They thought that Mike Weigold and Bruce Landsdowne and Gilbert Cook and others didn't know what they were doing. They said, 'You can't do that; you're not fisheries officers.' Saying that to lifetime fishermen—people who live and breathe fish!"

Its authority threatened, the DFO sent its big gun, the patrol vessel *Tanu*, to Viner Sound to confront these recalcitrant rednecks. The *Tanu*, 54 metres long, twin Fairbanks Morse thumping away, is an intimidating presence.

"We were all having a barbecue at the end of the day, and the wolves smelled it and were coming out of the woods," said Mike's wife, Lee. "Suddenly the *Tanu* came around the point. All of our boats had rifles on board, and without a word everyone got their guns out and started to take shots at the wolves." Faced with the reverberating crack of several 30-30s the *Tanu* turned sharply and was gone. The cat-and-mouse game had turned into open warfare.

For the next four years this group of fishermen worked at Charles Creek; they cleared the stream and, with the help of a biologist, planted eggs. "People risked a lot being involved in that," David Rahn said.

There were other similar incidents. For 30 years the fisheries department had discussed the rehabilitation of the Ahta River in Bond Sound, where early logging had destroyed the salmon habitat. It was with the agreement of the department then, that 15 fishermen set off in two seine boats to repair the watershed there. This time it was a

The 'Johnstone Strait Salmon Fisherman's Enhancement Society.'
Mike Weigold photo

provincial bureaucracy that bore down upon them. Officials from the Ministry of the Environment flew in and threatened the fishermen with charges of up to $125,000 for the first day of an offence. This time it was the DFO that came to the rescue.

"They brought the DFO in and they literally held their plane," Mike Berry said. "Finally they were told, 'Hold it. Go away. These guys have done a wonderful job.'"

Layers of bureaucracy have created an unwieldy and inefficient monster. "The division of responsibility between the province and the feds is getting harder and harder to deal with. That's one of the problems with the fish-farm issue," said Mike Berry.

Don Pepper is even blunter. "Never go to the fisheries with a problem," he said, "because they'll make a decision and it will seldom be the right one."

Recently the coho runs in the north—on the Nass and Skeena rivers—have been at record levels. Historically there have never been returns of this size. And the fishery is closed. "You have to wonder what's going on," David Rahn said. "The local managers on the ground deserve all the credit, and they must be tearing their hair out."

Don Pepper said, "Right now, off the coast, the pilchards have come back. There are 200,000 metric tons of sardines out there. Scientists say we can take 18,000 tonnes. For six years—the Americans believed they won World War II in less time—I've been trying to get a management plan out of the DFO. They have a new fishery here; they should have goals and objectives. They should be asking themselves, 'How many jobs could this create? How much landed value? What about economic development?' And then, most important of all, each year they should be accountable, should ask themselves, 'How have we done?'

"Last year they said, 'We're going to expand the sardine fishery. We want everybody's input. It's ITP—Invitation to Participate—time. Put in your plan for harvesting the fish.' They received 160 business plans. They said, 'We'll just go with the first seven.' And even then when the season opened they were still shuffling paper back in Ottawa; they weren't ready. It's just crazy. It's beyond comprehension."

The passive citizenry who let the 1957 spraying of the Nimpkish watershed go unchallenged have been replaced by people both knowledgeable and vociferous. As a result the bureaucrats in both levels of government find themselves faced with a barrage of criticism: outraged environmentalists, an unsympathetic media and warring biologists.

The biologists offer widely differing opinions. Whom to believe? Government biologists insist that everything is hunky-dory in the fish-farm business, for example. But Don Pepper said, "In the civil service a biologist doesn't have a lot of job opportunities, so he's loath to upset his superiors. And if he does there are an infinite number of ways to silence him."

"I've heard everything," said David Rahn, "from Dr. Dick Beamish of the Pacific Biological Station saying fish farming is low-risk to a fellow writing for the UN who says Atlantic salmon are the only species of salmon on the east coast because they've knocked out every other species."

There's no doubt where biologist Dr. John Volpe stands. He's an assistant professor in the Department of Biological Sciences at the University of Alberta, and he is studying the effects of the introduction of Atlantic salmon on the wild Pacific salmon stocks. He's tall and broad-shouldered with rumpled fair hair, and addressing a group at the Fisheries Centre at UBC, it's easy to see why he's *persona non grata* with the DFO. For starters he's the person who proved conclusively (contrary to DFO assurances) that Atlantic salmon had not only found their way into Vancouver Island rivers but were spawning there.

"There are all kinds of serious problems with aquaculture," he said. "If you look for them you'll find them, but the DFO prefers to look the other way." He noted that the use of antibiotics has created "superbugs," which are now appearing in the fish farms in Chile and Norway, that the pesticides used to control sea lice kill all crustaceans, and that the demand for fish food is consuming the marine resources of countries like Chile and Peru. Salmon are carnivores, he pointed out, and it takes 2.8 kilograms of wild aquatic species to produce one kilogram of farm food.

What concerns him most, however, is the quality of the food produced. He cited a 2002 Cornell University publication that mentions Canadian and Scottish studies published in recent issues of the environmental science journal *Chemosphere*. These studies report that levels of polychlorinated biphenyl (PCB) toxins are roughly 10 times higher in farm fish than in wild fish. They trace the source of the contamination back to commercial salmon feed. A recent feature in the British *Daily Mail* states that in Europe, "a chemical cocktail of substances found in trace amounts in these [farm] fish include canthaxanthin, a dietary additive which gives farmed salmon its appealing color; various pesticides such as cypermethrin, dichlorvos, and azamethiphos associated with cancer and reproductive problems in humans; copper and zinc-based paints; and malachite green, a fungicide." The paper notes that the latter was banned in June 2002 by the Scottish government and that canthaxanthin has long been banned by the EU for direct human consumption due to its links to vision damage.

The DFO maintains—not always successfully—that they are on top of things. Their recent report on sea lice is an example. According to John Volpe this report is so flawed that a group of scientists have asked the department's director of science to retract it. Otto Langer, a former DFO biologist and now the David Suzuki Foundation's director of marine conservation, criticizes it as being "unscientific." "It is more of a cover-up than science," he said. But given the DFO's impossible task of fulfilling their legislated mandate to protect wild salmon while at the same time embracing aquaculture, it's hardly surprising that it has adopted the dictum, "When in doubt, deny everything."

And then, as if all this controversy and conflict weren't enough, there is the puzzling problem of a fundamental change that scientists call a "regime shift." It's a fancy name for a pronounced change in weather patterns that was first noticed in the years 1976–77. Another one of these changes seems to have occurred in 1989. It's a complex phenomenon and only vaguely understood, but a key component is the Aleutian Low, which causes higher winds, pushes the warmer surface water away and brings colder water up from the ocean bottom. All of this impacts the marine species.

Of course, salmon have always gone through cycles of abundance and scarcity, the longest stretching over 30 years. And Steve Warren, who had lived longer than a lot of people, said, "I think this stuff about the salmon declining is misinformation. There were always tough times."

The First Salmon ceremony was proof of that. For thousands of years the First Nations people had welcomed each year's salmon run with reverence. Inherited wisdom told them that abundance was not guaranteed. For reasons mysterious and unknown the salmon sometimes came in record numbers and sometimes in disappointing ones.

Despite computer projections and DNA sampling and all the rest of the high-tech gadgetry, in some ways, as far as salmon are concerned, we are not a lot wiser than we were 10,000 years ago.

Consider Ron McLeod's "black box" theory. "Ron McLeod was a former director general of fisheries out here," David Rahn said. "He was an ardent conservationist and a knowledgeable man. He said that, basically, you did all this work, made sure the spawning grounds were okay, the eggs were fertilized, the fry were thriving, and then they just went into the black box and you hoped like hell they'd come back four years later. One little ship would go out and do a bit of tagging but you didn't really know what was happening. And now it turns out it's very complex out there."

"The fish are acting goofy," Don Pepper said. "They're spawning too early and they're 'getting lost'—things like that. Turns out there's a strong correlation between the water temperature in March at Kains Island and the diversion rate. If there's a strong El Niño and lots of warm water, it shoves the fish north. So you'll get Adams River sockeye in Prince Rupert."

Despite all man's meddling nature still maintains its mysteries.

C H A P T E R 1 1

High Boats:

Fisher Lassie — Thomas Hunt (BCP)
Nasoga — Chas. Wilson (ABC)

Pioneer Journal, August 23, 1950

In 1987 when the *May S.* was in Prince Rupert unloading fish, Spence Huson tripped on a hole in the sidewalk and suffered a concussion and a broken collarbone. The fall left him shaken psychologically as well as physically. He was 67 and had inherited his father's poor eyesight. Over the years he had become an extension of the *May S.* and his diminished vision was scarcely apparent; watching his seamless landings one would never have guessed that they depended more on instinct than on depth perception. But in unfamiliar surroundings things were different. "He realized that this was probably the end of fishing at his end of it," David said.

Fishing takes stamina and makes no allowances for infirmities. When he was 70, Sam Hunt said, "The worst part of fishing is old age. You can't cut the mustard. You can't put in the hours."

So Steve Beans phoned the family in Alert Bay to tell them what had happened, and Del Hall, who had fished with Spence for a dozen years, and the rest of the crew brought the *May S.* back to Alert Bay.

Then Spence Huson got his son and daughter together to discuss what was to become of his boat. He suggested giving the *May S.* to David and Grace, but as David said, it was hard enough for one owner to make a living let alone two. Besides, Grace's husband already had

OPPOSITE: *The* May S., *her paint and brightwork immaculate, at the Indian Breakwater in Alert Bay in 1996.* From the collection of Barrie McClung

MAY S
MAY S

his own boat. Instead, Spence sold the *May S.* to David interest-free. It meant a lot to David to have a boat of his own again, and it meant a lot to all three that the *May S.* was staying in the family.

The timing was perfect. David paid off the boat in four years. After the troubling years in the 1960s and '70s when the salmon seemed to be disappearing, they now returned in record numbers. Nineteen eighty-eight was a peak year; the Japanese market for salmon was booming and fishermen were getting seven and nine dollars a kilogram. For the next couple of years it was all good news.

"In 1990 I fished the 'blue line'," Don Pepper said, "and we caught 40,000 fish in one week. A huge run that year."

Despite the limited entry program, however, there were still over 500 fish boats chasing these fish, and in the 1980s the government continued their efforts to "reform" the fishery. They chose Peter Pearse to head a commission with a mandate to investigate every facet of the business. Peter Pearse was a handsome young academic with styled hair and good political connections, and in a matter of 11 months his preliminary recommendations were ready. Ignoring the depredations of industry, he asserted that the depletion of the fish was due to overfishing and that the answer was to reduce the fleet until "ultimately they might not need boats at all."

"Well, you see, the salmon resource drives economists crazy," Rod Hourston said. "They say, 'You could have a trap at the mouth of the Fraser River. Let what you don't need through. That's all you need if you want to maximize the efficiency of the resource.' And of course that's true. But we're talking about a way of life here. Some people fish to live but a lot more live to fish. Twenty per cent of the people catch 80 per cent of the fish, but still they're all out there. 'Just one more season.' 'I'm going to make it this year.'"

In the sweeping changes Pearse proposed there was no calculation of the cost of welfare for a whole segment of the population presently earning its way, if modestly, and living in hope of that next good season. Instead Pearse argued that salmon should not be "common property" at all, but the property of those enterprises that could use the resource most effectively.

Changes in government from Liberal to Tory and back again altered this philosophy not at all. By 1996 Fred Mifflin, a former admiral, was Minister of Fisheries and Oceans in the Liberal government. He announced the Salmon Revitalization Plan, which aimed for a 50 per cent reduction in the capacity of the commercial salmon fleet, single gear and area licensing, and licence stacking. As before, there were no plans for conservation or enhancement of the fishery—and no plan for the people who would have their livelihoods and communities destroyed.

There were cries from the sinking ship. In Ottawa in October 1997 Senator Pat Carney stood up and read out a letter from Andy Erasmus of Masset in the Queen Charlotte Islands that said, in part, "Many people who had never been out of work in their lives are now forced on the dole." Another letter, from French Creek on Vancouver Island, said, "It is a no-win situation. We can't afford to get out of [the fishery], but we are not able to access credit to stay in it. Please help us and others that are in this dilemma."

But the cries went unanswered, and one by one the little coastal communities, Alert Bay among them, went under. There was much talk of retraining plans, and money was thrown at something called "short-term crisis management," but in the end it was every man for himself, every woman for herself. Mike Weigold and his wife, Lee, were among those men and women. Mike was judged by people like David Rahn and David Huson to be a matchless fisherman.

"He was our sonar man," David Huson said. "He could look at the sonar and tell you everything about those fish."

"Mike had an uncanny fish sense," said David Rahn. "He could stand on the breakwater, look out and find a jumper we didn't even see, tell you what kind of salmon it was and which way it was going."

"Being able to see fish from a distance is something you need to be trained for," said Mike. "One of my first jobs was with Byron Wright on the *BC Maid* when I was 13. Byron paid me $20 a week plus five cents a jumper, provided I could describe type and direction. He docked me a nickel for any profanity. Well, one day we wandered into Wells Pass scouting for fish and ran into the motherlode—jumps everywhere. Byron finally cut me off, but he kept billing me for swearing!

"Seining is a funny business," he went on. "A large part of what you are doing is trying to figure out when the fish are coming, where they are coming from, who is 'on them' and who is watching. There are fishermen who are above average in the area they work. So you have a choice: figure out how the fish run in an area you've never been before, or follow someone who does have it figured out. Then you have two options; you can either jockey for position or just follow him around."

David Rahn adopted the latter course of action. "I was part of the 'elastic-band' fleet," he said. "When Mike took off in one direction, the rest of us followed as if we were connected by an elastic band."

But in 1998 Mike left the fishery he had entered with such enthusiasm and energy. He stopped doing what he did so well. He sold the *Bruce Luck* and his licence. He was out of fishing for good. "I had no faith in the DFO's ability to manage the industry," he said.

The thrust of the federal government's fisheries plans had always been the reduction of the fishing fleet. There was no corresponding effort to enhance or protect the resource itself, and the Mifflin Plan was no different from its predecessors in this regard. Forced to the wall by academics and admirals, the people of the Broughton Archipelago, like many others on the coast, resorted to do-it-yourself.

In 1978, four years before Mike Weigold and his disenchanted colleagues set sail for Viner Sound, Billy Proctor and some friends decided to rebuild the Scott Cove salmon run. Neighbours in Echo Bay, and Gary Ordano and his skidder in Scott Cove, pitched in. Billy Proctor has spent his entire life among the islands of the Broughton Archipelago and he's alarmed and angered by the damage that is being done to salmon stocks. Encouraged by the local fisheries warden, Bill McLeod, assisted by DFO and financed by an American sports fisherman, of all things, he and his group worked to clear the creek bed of debris and build a hatchery. For almost 25 years a small army of volunteers, becoming more expert year by year, have revived the local stock and built up the run in Scott Cove to some 4,000 spawners.

In April 2002 Stephen Hume, writing in the *Vancouver Sun*, raged at the destruction logging has caused to the waterways that provide habitat for salmon. He saw the Fisheries Act being ignored and he proved his point with a picture of Clayoquot Sound, where logging has stripped a valley bare and left a naked river without a vestige of cover. But in the Nimpkish River valley wondrous things were happening. With funding from Forest Renewal BC (FRBC), an unlikely alliance was attempting to right some past wrongs.

Using money from stumpage fees, FRBC initiated habitat-restoration projects, one of which involved the Nimpkish River watershed. As early as 1984 a group called the Nimpkish Resource Management Board was formed. It consists of members from Canfor, the 'Namgis First Nation band, the IWA and TimberWest—all the stakeholders in the valley. Since 1995 technical working groups from this board have met on a regular basis to oversee the rehabilitation of the river. There are now in excess of 750 individual site treatments in the valley, and each summer 10 to 15 people—half of them from Canfor and half from the 'Namgis First Nation—work on stream restoration for eight to 10 weeks. They are an enthusiastic work force. "The crews would do this all year if they could," said Charlie Jancsik, the project co-ordinator. Instead, virtually all the Native members have been able to go on to jobs with Canfor.

To date, Mike Berry estimates, FRBC has spent over $6 million to repair the damage wrought by 50 years of logging. But how much longer this project will continue is debatable. Forest Renewal BC is now the Forest Investment Account, and although its mandate remains the same, its funding is now doled out year by year and may dry up completely because of the present government's commitment to fish farming. The same uncertainty faces the hatchery that the 'Namgis band, with financial and technical assistance from DFO, has established on the river.

Fish farms first moved north of Seymour Narrows to the area adjacent to Johnstone Strait in the late 1980s. Even in good seasons they were competition; in bad seasons they were the kiss of death. "Prior to this there was a predictable market demand," David Rahn said. "High

Mike Weigold, considered a top fisherman by his peers, sold his boat and licence and left fishing for good in 1998. Mike Weigold photo

prices at the beginning of the season and then they tapered off as more fish were landed. In a poor season the price would stay high because of the scarcity, so sometimes fishermen made just as much as they did when the fish were more plentiful." But now there was an economic dislocation because of the farm fish. "The awful thing is, now you're getting poor seasons and poor prices."

Those people with crappy old boats couldn't make any money—but they had a salmon licence. "So what would you do?" Don Pepper asked. "Keep fishing a couple of days a week in season or sell the licence? You take the money and then you're out of it. And by the time you pay the tax, you haven't got enough to do anything. There are four seine boats fishing out of Alert Bay now. Four boats. It's a village with a long, long fishing tradition and it's the only thing they've got."

Even the village's Native food fishery is resented. The salmon that the village of Cheslakee caught as food for the winter became the salmon that the First Nations people shared with hundreds of other commercial fishermen. Then they became the salmon that were scarce and claimed by a privileged few. Now an Alert Bay boat that tries to set its seine off Double Bay is encircled and harassed by sport-fishing boats that race over from the Plumper Islands and surround it, thwarting its every attempt to make a set or even get its net out of the water.

"It's quite a business now," said David Huson, who was faced with tougher markets. "You're trying to get quality."

To get quality he spent a winter and some $70,000 putting tanks in the *May S.* Mike Weigold, who did the same thing with the *Bruce Luck*, described the process. "If you just have to make partitions and do the foam and glassing it's not such a big deal—maybe $20,000 or $30,000 and a month's work. But when the boats are older, and wooden, the time goes up—and the cost. You can't glass over rot. The bulkheads and the shaft log have to be watertight, and any machinery or hydraulic hoses have to be moved. Both David and I had to enlarge the hatch openings to access four tanks. You can split the hatch in two or three but that's not such a good idea—too much free-surface effect. The fish are suspended in liquid and ice. The boat lists and the liquid moves; the boat lists more and the next thing you know you're walking on the keel." The hatch covers, like the tanks, were lined with foam and covered with fibreglassed plywood for durability.

The tank system wasn't a new idea. As early as August 1963 the BC Packers boat *Nahmint* arrived in Alert Bay using this experimental method of packing fish. The *Nahmint* is remembered not so much for this new process as for the fact that on landing at the BC Packers' dock her reverse gear failed to engage and she snapped off two pilings, buckled the plank walk and set two floats adrift.

By the 1980s, tanking a boat was common practice. Instead of having crew icing fish manually, which was a slow laborious job, the hold was commonly divided into four tanks, at least two of which were filled with ice. Salt water was added to create "a nice slurpy," and as the fish were caught, they were dumped down a manhole into this mixture of ice and water. A boat like the *Bruce Luck* would usually load a tonne of fish in each of these two tanks. Once the tanks were full it took a long time for the temperature to change, but if it began to rise after three or four hours pumps circulated the water and cooled it down again. This system meant that the fish were loaded more quickly and were less likely to be bruised and battered by pughs and the occupants of heavy waders. But its greatest benefit became apparent when the seiner got

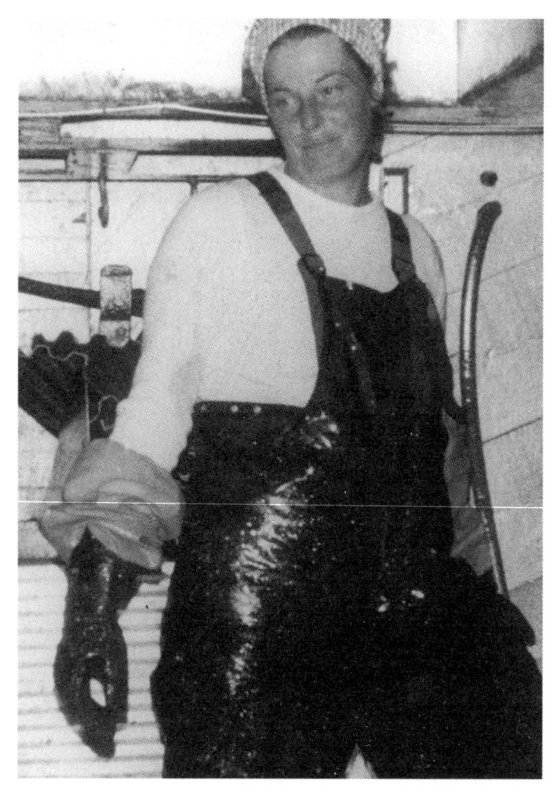

An exhausted Lee Weigold delivering pink salmon in 1983. Mike Weigold photo

back to the plant or the packer. A pump could suck out the unblemished fish in something like 30 minutes as opposed to four or five hours of "hand bombing."

This was clearly a new and better way of handling fish. The real news, however, still concerned the fish farms. As early as the 1970s Ian Todd remembered the DFO talking about allowing the import of Atlantic salmon eggs from Norway. "I couldn't *believe* what I was hearing," Todd said. "It was already known that things had gone very wrong in Norway; they had got to the point where they were using rotenone to kill everything in their streams to eradicate disease. I have long thought, in my dinosaur approach to life, that we should look after what we have. Bring the logging lobby in by the nose and make them clean up their damage." For a while, on the Nimpkish River, that was happening.

The weather held. The next morning David and Barrie discussed their options for the day.

"Let's take a run over to Gilford Island," David said. "You okay with that? If you can check the oil and the clutch oil." He flipped the last of his cigarette over the side and stood over the hatch to the engine room.

"Everythin' good?" he shouted.

"Yep. I was just going to check the batteries."

"They should be good. I just spent over two grand on them last year and Larry checks them all the time for me. He's pretty good, you know. I mean he's pretty good with machinery. You remember him from the Bay. He used to own that little gillnetter *Seal* with that Chrysler Crown. Remember we got stuck in Sointula and it was a crack in the distributor cap? He's got a really nice boat now. Okay to fire her up?"

The Volvo roared to life.

Broughton Archipelago is a fancy name that would make the old-timers snort. It refers to the mass of islands crammed between Johnstone Strait and the deeply indented coastline of the mainland. Gilford, the largest of these islands, is big enough to contain several lakes and streams and its own respectably high mountains. Circling it like a car in a traffic roundabout, a boat can veer off in any number of directions—into Bond Sound, Thompson Sound, Knight Inlet, Clio Channel and Chatham Channel. These exits vary widely in breadth and length, and nearly all their shorelines, as well as those of Gilford Island itself, contain the sites of ancient Native villages, some still used.

"Maybe go up to Simoom Sound where the old man took us goose huntin' that time," said David. "You remember that?"

Barrie would never forget it. "'We'll go goose huntin',' your dad said. Your uncle loaned me a double-barrelled shotgun and showed me how to load it. Just a massive big gun I could barely lift. I was so excited I nearly pissed my pants."

"Took off in the skiff," David said, "me rowing, my dad in the stern and you in the bow. 'Now when you come onto some ducks,' my dad said, 'you give them about a six-foot lead.'"

"I didn't know what a six-foot lead was."

"You didn't know how to shoot either, but I told ya." David was laughing.

"Told me to get on the front trigger and give it a pull, you bastard."

"I couldn't resist it." David was wiping his eyes. "I couldn't resist it. I had the same trick played on me."

"So I hit both triggers and both barrels went off and it threw me into the bottom of the boat." Now Barrie was laughing, too.

"Just about deafened us all."

The *May S.* slid by a bluff. There were pillows of moss clinging to its granite face and they were brown now at the end of this long warm summer. A few small trees, twisted like bonsai, gripped the rock and drove their roots into its nutrient-poor crevices. At the top of the bluff, banked by salal, Barrie caught a glimpse of a piece of plywood cut into a stylized shape and painted with a Native design.

"What was that up there—that painting on top of the bluff?"

David looked away, almost embarrassed. "It's where a kid, ya know, jumped in. Committed suicide. They do that now. They're not workin'. Get into drugs."

Barrie stared at the black water that raced past the base of the bluff and tried to imagine deliberately jumping into the tide. "We never heard of suicide when we were kids here."

"They have nowhere to go," said David.

They had the chart spread out. Bond Sound, the most easterly exit, is a short, serenely beautiful run that ends where the broad Ahta Valley dips down to the sea.

"It's really pretty up in Bond Sound," said David. He's spent his life in pretty places. His favourite is farther north. "Gunboat Pass is really something though," he told Barrie. "Gunboat Pass is beautiful."

Knight Inlet, on the other hand, is pure drama. It twists for 113 kilometres deep into the Coast Mountains until finally blocked by the bulk of Mt. Blair and Mt. Waddington. All along its length mountains wall it in and waterfalls drop from their heights. The waters of the inlet change from the clear black depths at its mouth to the bright unnatural turquoise of glacial runoff at its head. The Klinaklini River skirts Mt. Blair at the head of the inlet and here the eulachon run in spring. There are the remains of a large and ancient seasonal village site here, where Native people from the coast and the Interior gathered to trade in eulachon oil and where the coastal First Nations people still go to catch these little fish and turn them into oil.

"There isn't much left up there now," David said. "And they're loggin' at the head of the inlet and pushin' through roads. But at Dzawadi they still make eulachon oil in the spring. It's quite an undertakin'—so much work and it's complicated. I mean, the temperature has to be just right. They take the kids up there to learn how to make it. The young people need that."

"I was up there years ago with Steve," Barrie said. "You had to anchor about a mile out because of the shallows at the mouth of the river. We shoved the *Kitgora*'s nose in tight against the bluff and let that waterfall pour down on the front deck to cleanse our souls." He chuckled. "The only thing that got cleansed that day was my sleeping bag and mattress because nobody latched the skylight."

David looked into his coffee mug and reviewed the stone cold contents. Then with a deft flick of the wrist he fired the remains overboard. "I always wanted to take the *May S.* up Knight Inlet one more time and get a photo of her with her bow under that waterfall. It's kind of a spiritual place. And you've got that good camera ..." But the *May S.* didn't alter her course. Instead she glided across the entrance to Knight Inlet and went on.

"Changin' the subject," said David, "where do ya think we should stop tonight? I mean I don't want to be lookin' for a place after dark. We could just run through here and let the hook go at Port Harvey, eh?"

Chatham Channel, in contrast to Knight Inlet's broad, deep course, is a tortuous little passage, narrow and shallow. The *May S.* plowed along, passing Minstrel Island, and entered the funnel opening of this channel. Suddenly the depth sounder jumped from 75 metres to 35 and then to 20 and then to 12. Just under the surface of the water long fronds of chestnut kelp, like the rippling hair of 1,000 mermaids, streamed away in the tide. There were range markers ahead; David brought the *May S.* carefully around until they lined up one above the other.

"Can you get through here on any tide?" said Barrie, viewing the kelp with some distrust.

"Well, on a zero tide you'd be pretty touchy. You wouldn't go roarin' through here. And they've got lights but you wouldn't like to go through here at night. We've done it with a load of herrin' but the tide has been right and that."

They were in Havannah Channel now. "That's Matalpi village over there," David said. "The old name is Etsikin. That's where they had the good clams. I used to buy clams there. My dad and I dug clams all over the place when I was 10 or 12. We used to go through all the little passes. When we went out huntin' we'd stop if we went by a clam beach and get some clams to take home. If you're just travellin' around on the boat, like."

Everywhere on their journey—in every quiet bay, appearing around every corner—they had encountered floating checkerboard patterns on the surface of the water. They looked like nothing so much as partially submerged tennis courts and to Barrie they were a surprise.

"Fish farms," David said. "They're doing' this all over the coast now and everybody's protestin'."

It was dusk when the *May S.* rounded a point, slowed and drifted. David, on the foredeck, readied the anchor.

"Kick her into neutral," he hollered. The *May S.* glided to a halt and there was a splash as the anchor dropped into the water followed by the rattle of chain. A kingfisher, disturbed by the noise, set up a raucous chattering and darted across the water to land on an overhanging hemlock. Then the engine stopped and there was silence.

"It's a shitty bottom here," David said. "Got my anchor snagged in a rock crevice here a few years ago with the *Miss Donna*. Had to call the old man on the radio to give us a hand. He had his boom over the side pulling with his winch and I was pulling. Damn near pulled his boat over before it let go."

In the still darkness the crew of the *May S.* leaned on her rails and surveyed the sheltered bay that was Port Harvey. Above the mast great swaths of brilliant stars arced over them, and around them the black water lay immobile. There was no chuckle and slap of water against the hull—no sound at all—until Barrie suddenly jerked to attention. "There's jumpers out there, David. Hear that? I hear jumpers."

"It's the fish farms," David said. "Some of them salmon get out and they got no sense of direction, eh. They just jump around." He dragged on his cigarette and its tip glowed brightly in the dark. "Like I say, we used to dig clams here but I don't think anybody does it any more because of the fish farms. We just passed three of them back there. Every little bay's got a fish farm and they kill everything on the bottom.

"I went into Save-On-Foods three or four years ago," he said, "and talked to the fella that buys their fish. They buy all farm fish, eh. They order one day and it's delivered the next, all packaged and ready for the shelf. It doesn't matter if the weather's too tough to fish or what. There it is next day. And the price they pay—I wouldn't even be able to pay the crew for that money."

"Will they take over, do you think?"

"Looks like that's going to happen. Next generation, I guess it will just be 'salmon.' Nobody'll know the difference."

"Next generation it'll be mad salmon disease," said Barrie.

CHAPTER 12

High Boats:
Ha-Wha-Las — George Alfred (BCP)
Ermalina — Louis Benedict (ABC)

Pioneer Journal, September 20, 1951

In 1988 Howard White published a wildly funny short story called "Morts" that helped him win the Stephen Leacock Memorial Medal for Humour. It concerns the disposal of a truckload of dead fish from a Sunshine Coast fish farm. Maybe 1988 was the last year anyone could write something funny about the fish-farming business. For even then, 17 years after the first fish farm was established in BC, the industry was still the centre of a raging controversy.

In the area around Alert Bay it wasn't an issue yet because the majority of the farms were on the southern coast in and around the Sechelt Peninsula. But the water there is warmer than it is north of Seymour Narrows, and Howard White's truckload of dead fish attested to the fact that it isn't the ideal location for fish farming. So in the late 1980s the farms started to move north.

Many of the small fish farmers had exhausted their capital at this point and went bankrupt. Huge conglomerates promptly bought them out. Stolt, a Norwegian company, and Heritage Salmon, owned by BC Packers (now part of George Weston Ltd., the giant food company), soon held a monopoly on the aquaculture business. The coast's inhabitants thought uneasily of the parallel with the logging industry, where a few huge corporations claimed the forests as their own.

By 1995 the public's anxiety over the environmental consequences of fish farms had caused an outcry that the governments couldn't ignore. The T. Buck Suzuki Environmental Foundation, the Georgia Strait Alliance, the United Fishermen and Allied Workers' Union, Greenpeace, the Sierra Club of BC, the Steelhead Society of BC and private individuals like whale researcher Alexandra Morton demanded an investigation of the industry.

Author, photographer and environmentalist Alexandra Morton is a relentless critic of fish farming on the West Coast. Kristine Larsen photo

The BC government was forced to respond; they placed a moratorium on new farm tenures and commissioned a Salmon Aquaculture Review. The five issues that the review was to consider included the impacts of escaped farm salmon on wild stocks, disease in farm and wild fish, environmental impact of fish-farm wastes, the impacts of the farms on coastal mammals and the siting of the farms themselves.

The Environmental Assessment Office established a technical advisory team of experts charged with preparing a report and making recommendations, and the resulting review was comprehensive and clear. Although the team advised that in their opinion the salmon aquaculture industry presented a low probability of risk to the environment and to humans, it also stated that there were significant gaps in the scientific knowledge on which their conclusions were based. Furthermore, they advised that "even though the risk of significant environmental impacts has been determined to be low, the possibility nevertheless exists for the occurrence of a catastrophic event." Therefore, over and over in the report, they stressed the importance of "rigorously enforcing the implementation of standards." The report made 49 recommendations and suggested that "in the absence of clear standards [and] strict enforcement ... suspicion remains high and strong criticism continues."

The government, with the peculiar logic characteristic of bureaucracy, gave the impression that by producing this pile of paper they had more or less solved the problems connected with fish farms. They appeared to dismiss the warnings issued by the Salmon Aquaculture Review, a study they themselves had commissioned, and as a result, the "suspicion" and "criticism" mentioned by the review remained—and with good reason. The issue of escaped farm salmon was a good example. In 1984 when Atlantic salmon were first imported to BC, there was a series of assurances from the provincial and federal governments, and from the industry itself, that the salmon wouldn't be able to escape. And if they did escape they wouldn't survive or reproduce. But an Aqua-Vision Consulting report prepared for the BC Salmon Farmers Association in July 2001 confirmed that, contrary to these earlier assurances, escaped Atlantic salmon had successfully spawned in BC. The industry's fallback position was that "it remains to be determined if a self-sustaining population can be formed."

Perhaps one of the most unsettling pieces of information on the subject was a memo that Alexandra Morton quoted. It was dated February 19, 1991, some five years after the first farms were established here, and it was written by Ron Ginetz, federal chief of the aquaculture division. Despite the problems that surfaced in Norway in the 1970s, Ginetz remained an enthusiastic proponent of fish farming in BC. But his memo warned, "In my view it is only a matter of time before we discover that Atlantics are gaining a foothold in BC. Do we prepare the public/user groups for the possibility, and strategically plant the seed now, or do we downplay the idea and deal with the situation if and when it occurs?"

Credibility, therefore, is in short supply, and people like Morton continue to write "10,000 letters" to government ministers expressing their mounting anxiety. Alexandra Morton lives in Echo Bay. She's an award-winning author and photographer, and her whale research has been published in prestigious international scientific journals and documented by National Geographic, ABC Television and the Discovery Channel.

Alexandra's fascination with orcas brought her to the Johnstone

Strait area in 1979. In the following year's field season she met her future husband, Robin, a man who shared her passion for the marine environment. They bought a former police launch, which provided mobility and a home for a family that now included a baby son, and they set out to see if recording the orcas on film and in print could make them a living. And then, one day, doing underwater filming at Robson Bight, Robin drowned. There on the boat with her small son and the horrible suspicion that her husband's body lay somewhere in the water beneath her, Alexandra experienced the trauma that is the non-negotiable price nature so often exacts.

This wild and beautiful coast has never been a place for the faint of heart. For a woman alone it poses enormous challenges. So people expected that Alexandra Morton would turn and flee. Instead, irrevocably hooked by the splendour of a landscape she had come to love deeply, and by the freedom it afforded, she dug in her heels and survived—sometimes only just. Her decision to make a life for herself at Echo Bay required plenty of old-fashioned grit; now, 20 years on, having proven her right to be part of this remote world, she is outraged at what she perceives as the threat of fish farms. "Living here," she said, "I am witness to the death of an archipelago."

For 12 years she has tracked the diseases, parasites and chemicals that have appeared along with the fish farms and has joined others in the area in a futile attempt to get the DFO involved. Morton is concerned about the use of the lethal pesticide, ivermectin, to combat fish lice and the use of an equally toxic anti-fouling paint, Flexgar, to coat net pens. In a letter to Allen Rock, the minister of health, she quoted the label on the Flexgar container, which reads, "Toxic to aquatic organisms. Do not contaminate water. Do not allow chips to enter water."

Others worry about the taxpayer's liability. Everywhere in the world that salmon farming has taken place, epidemics have occurred. In many locations governments have paid out millions of dollars to fish farmers in an attempt to compensate them for their losses. In 1999 the industry publication *Intrafish* reported that in New Brunswick losses totalled $50 million and a good part of that cost was paid out in compensation by the provincial government.

Despite the controversy over potential risks to wild salmon stocks, salmon farms continue to proliferate on the BC coast with the enthusiastic backing of the provincial government. Don Millerd photo

And the epidemics themselves are a major concern. Already there have been outbreaks of IHN (infectious hematopoietic necrosis) and furunculosis at farms in the Broughton area; the former disease resulted in millions of dollars' worth of salmon being destroyed and the latter required the use of the antibiotic erythromycin. A DFO publication contains the following warning: "Fish treated with erythromycin, including spent adults, may not be used for human consumption and under no circumstances may the drug be used orally … in food fish."

In 2000 Canada's auditor general joined the crusade. "A major constraint to enforcing habitat provisions," he wrote, "is the Department's lack of scientific information that would enable it to develop administrative criteria for what constitutes harmful alteration, disruption or destruction (HADD) of habitat with respect to salmon farming. Without such information, field officers do not know how to monitor farming activities to ascertain compliance with the act. In summary, we have concluded that the Department (of Fisheries and Oceans) is not fully meeting its legislative obligations under the Fisheries Act to protect wild Pacific salmon stock."

In November of that same year the David Suzuki Foundation commissioned retired Supreme Court justice Stuart Leggatt, a former MP and MLA, to inquire into the impact salmon farming was having on the BC coast. No representative of the BC Salmon Farmers Association attended the inquiry; the participants, according to a spokesperson, "lacked the scientific qualifications necessary to present valid opinions." The association was right about scientific qualifications; hardly any of the presenters had any. Certainly Chief Bill Cranmer, chair of the Musgamagw Tsawataineuk Tribal Council of Alert Bay, didn't have any. All he had was apprehension. "Our greatest fear," he told the inquiry, "is that the wild stock will be destroyed."

Neither level of government participated in the Leggatt inquiry, either. "While the government of Canada certainly welcomes this gathering of public opinion on BC salmon farms," said federal minister Herb Dhaliwal, "we will not be participating."

John van Dongen, the provincial Minister of Agriculture, Food and Fisheries, said, "We are already addressing the environmental issues facing the industry." If so, one wonders, why not share that information with a concerned citizenry? And why turn your back on an opportunity to follow exactly the Salmon Aquaculture Review's recommendation that "public input is essential, not only to encourage well-informed decisions and reduce conflict, but as a matter of fairness"?

And then in February 2002 the provincial government announced that it would lift the moratorium on fish farming as of April 30 and that the industry would become "self-regulating." The response to this announcement was charged with emotion.

The environmental and aboriginal representatives on the province's salmon farm advisory group walked out in disgust, feeling they had been betrayed.

The governor of Alaska appealed to BC's Ministry of Agriculture, Food and Fisheries. The governor has a lot to lose, and escaped Atlantic salmon in Alaskan waters are just one of his worries. The thriving wild fishery of his state will be devastated by competition from farm salmon. For it's not only disease that threatens the wild stock, but also market forces. Fish that can be sold for under five dollars a kilogram wholesale, and that can be sold fresh year-round in local markets or frozen for sale overseas, could destroy the market for the wild product. Already Alaska canned salmon are selling for fire-sale prices in the supermarkets.

Rafe Mair, then the top-rated talk-show host in Vancouver, devoted two weeks and an impassioned editorial to the issue. And the editor at the *Vancouver Sun* wrote a long and thoughtful editorial headed, "Salmon farming decision unleashes a juggernaut."

Anita Peterson of the BC Salmon Farmers Association thinks there are a couple of reasons for this clamorous response. "For one thing," she says, "people are more environmentally aware than they were and more aware of what they are putting in their mouths in this day of genetically modified foods. And we're the new kid on the block so it will be a while before they trust us."

The second reason for this emotional outpouring, Peterson thinks, is that fishing is not a job, it's a way of life. She knows. She is another who has faced the challenges of the coast as a woman on her own. She was a fisherwoman herself for nine years.

The BC Salmon Farmers Association works hard to counter the public's criticism of the industry. They point out that the Salmon Assessment Review was one of the most rigorous and costly in the history of the province, and that "since October of 1999 the government has worked toward implementing policies that will result in the most comprehensive regime of any jurisdiction in the world." The association has also created its own 23-page Code of Practice, which outlines, in great detail, standards for every facet of the industry.

Anita Peterson says these standards aim to create an agricultural enterprise with minimum impact on the environment. The salmon farmers argue that they are the most rigidly controlled type of agriculture in the country. They point out that the feedlot that contaminated the water supply in Walkerton, Ontario, and the runoff from pesticides and sewage that seeps into rivers and oceans every day from land-based farm operations, make fish farming an exemplary form of food production.

In a concise table published by the provincial Ministry of Agriculture, Food and Fisheries in January 2002, the government outlines the recommendations of the Salmon Aquaculture Review, the actions taken to date and the actions to be taken in the future. Cutting through a tangle of bureaucracy, it's an impressive attempt at clarity.

Unfortunately, there's a missing factor so important that it negates all the effort put into these documents—and that factor is enforcement. Although there will be audits, much of the monitoring process is to be done by the fish farmers themselves. As mentioned, it is to be a "self-regulating industry." But if you believe that a business carried on in a remote location, with no real supervision and no real penalties, won't bend or break any number of elaborate rules if it's profitable to do so, then it's a good bet that you also believe in the tooth fairy.

The Fisheries Act's aquaculture regulations state that, "No person shall provide false information." Well, yes, but what happens to someone who does? Very little, it appears. The fines for allowing farm salmon to escape are $2,000, too small to be of any consequence. Other instances of non-compliance are enforced by a graduated scale of measures that "may include written warnings, violation tickets, administrative penalties, formal prosecution and land tenure/aquaculture licence suspension/cancellation." What irreparable damage may be done while following this laborious process to its conclusion? Kirk Stinchcombe, the assistant manager of salmon aquaculture, said that this is not an issue because of "the assumption that salmon farming presents a low level of risk." Presented with that quote from the Salmon Aquaculture Review—the government's own study—about the possibility of "a catastrophic event," he abruptly changed the subject.

Worse still, because of its conflicting mandates, the DFO doesn't appear to be enforcing its own Fisheries Act. "The Fisheries Act," stated Jim Fulton, executive director of the David Suzuki Foundation, "is the strongest piece of environmental legislation in this country. It was written to protect the marine environment and wild fish stocks." And yet Stephen Hume's article in the *Vancouver Sun* was headlined, "Failure to enforce Fisheries Act could have hidden consequences," and the photograph accompanying the article shows a river running through a denuded valley.

The points that the Salmon Aquaculture Review made over and over again, that "a comprehensive surveillance program is essential" and that standards must be "rigorously enforced," have been ignored. The governments, it appears, are once again offering the "sympathetic administration" that was extended to the logging industry.

Tucked away in UBC Library's Special Collections there is a two-volume manuscript of Henry Doyle's memoirs. Like virtually every retiree on the face of the earth, Henry Doyle planned to write a book; unlike most other retirees he actually did—and a very long one at that. He wrote clearly and well and had some important information to leave behind.

Henry Doyle was an American who came to Canada in 1895 to manage and eventually inherit his father's fish business. He spent over 60 years in the industry and was appointed the first manager of the British Columbia Packers' Association. He seems to have been a man confident of his capabilities yet without ego. He called a spade a spade, identifying the men of integrity he had encountered in his lifetime and also the scoundrels and nincompoops (which may have been why his manuscript was never published). He voiced the opinion, backed by research, that salmon runs have always been erratic and fragile due to the vagaries of nature and that the destruction of habitat can only exacerbate this. He pointed to the demands of agriculture on waterways, and the damage that logging and hydroelectric dams have done, and passionately urged the governments to address the problems before it was too late. This, of course, was written 50 years ago and the pressures on our waterways have only increased.

Ian Todd has never heard of Henry Doyle but he shares his opinion. "I get annoyed because people say the salmon are declining because of overfishing," said Todd. "Controlling the commercial fishing is the easiest part of it, because in the end we can shut it down if it's necessary. But is there any point in doing that if the fish have nothing to spawn in but a sewer pipe, or if they get sucked into irrigation systems? It's the pulp mills, the dams, the logging practices and the increase in the human population that have brought the salmon to this state."

In 1952 an editorial in the *Province* newspaper was entitled "The Great God Hydro;" it outlined the continuing friction between fisheries and hydroelectric interests over the use of BC waterways. At about this same time, Henry Doyle wrote that fresh water was the critical issue for the fisheries. He was weighing the claims of agriculture and logging against those of the fishing industry. He couldn't have envisioned fish farms as the answer to this dilemma. But for governments faced with our voracious demands for lumber and power and employment, perhaps fish farms are an answer. As Alexandra Morton told the Leggatt Inquiry, "Along comes a salmon that does not need a river and ... suddenly to a politician, to someone who is sitting in an office, they are thinking, 'Okay, we can have salmon and we can destroy the rivers.'"

The multinational aquaculture companies, of course, think that farming salmon is an inevitable development in the ongoing need to supply increasing populations with good fresh food. Don Millerd thinks that, too. The Millerd family have been in the fish business for three generations. It was Don Millerd's grandfather, along with a partner, who formed the Gosse Millerd Packing Co. In the 1920s the company had a number of canneries on the coast, and Don Millerd grew up in the business.

The Husons fished for the Millerds for almost 25 years. In the late 1940s a mariner following in the wake of the weather-worn little fish packer that chugged its way down Johnstone Strait might have been understandably surprised to find that the helmsman was a helmswoman—a grey-haired woman at that. It wouldn't have surprised the Husons, of course, because the woman piloting the *Shumahalt* was David's grandmother. She had no intention of letting her husband's poor eyesight—or virtual blindness, let's face it—deprive them of their living, and continued to guide their boat on the 22-hour trip to Vancouver, where the catch was delivered to the F. Millerd & Co. cannery.

Don Millerd spent his summers working on packers and dispatch boats and in fish camps, and he spent his winters getting an education. He had graduated in law and was articling before he knew, beyond all doubt, that he didn't want to be a lawyer. The north coast had seeped into his soul and he had no intention of leaving it for a suit and tie. What he really wanted was to work in the fish business. So in the 1970s he borrowed money, scrounged equipment and started his own cannery in North Vancouver. For a new business the first few years are critical, but Millerd's not only survived, it flourished. By the early 1980s he felt like a man with a tiger by the tail. Salmon were reasonably abundant and the markets were strong. He had a processing barge in Sointula on Malcolm Island, where some of the salmon were dressed and then trucked to the city to be frozen and sold to Japan. He had fishermen working for him out of Alert Bay and Sointula, and he was doing what he had always wanted to do.

And then in 1984 two things happened: interest rates rose until Millerd's business was paying 24 per cent on borrowed money, and over in Belgium a man died from botulism contracted from a can of Pacific salmon. Overnight the Europeans stopped buying canned salmon and supermarkets cleared their shelves of it. And overnight Don Millerd went bankrupt. It was astonishing how fast it happened, he recalled.

In the ensuing months, in order to survive, he arranged with the court-appointed receiver to rent his own property so that he could still carry on the business in a greatly reduced way. It was a hand-to-mouth operation and a humiliating one. He was required to pay cash for everything he bought and was reduced to collecting his receivables (if he could) before they were due in order to pay his bills. "I was working day and night," he said. "I was so stressed I could hardly string words together to make a sentence."

But the mind is a curious thing. Somehow, through all these very immediate troubles, he kept thinking back to a visit he had made to Denmark a year or two before. Millerd sold chum salmon to an old fellow named Shell Angley who had a smoking plant there. On this visit the Dane said, "I want to show you something," and took him into the plant where they were processing salmon. They were nice chums, Millerd recalled, but the odd one had a bruise or some blood spots. The workers were trimming them.

Then Angley said, "Now look at this." He took him to another table where other fish were being readied for the smoker. But every one of these fish was perfect. There were no bruises, no bloodspots.

Don Millerd, a self-confessed "fish guy," couldn't at first identify the fish at the second table.

"They're Atlantic salmon," Angley said. "They're farm fish."

"I was looking at that fish," said Millerd. "It was firm. It was nice. I thought, 'Wow, if this works out it's going to be awesome competition. I'd better watch out for this.'"

From that moment on Millerd paid attention to the fish-farm industry. In 1985, a year after his company went into receivership, the first fish farms in BC were being established in the Sechelt Inlet area.

"Jesus, this is going to work," said Millerd. He was struggling to survive and he had a cannery that had nothing to do in the winter. He wanted to get involved.

His first attempts could hardly have been termed a success. He got a big seiner and he and his crew filled the hold with water and fish. Then they got an oxygen tank from a welding shop and started for Vancouver, bubbling oxygen into the hold as they went. The tank only lasted minutes and the fish not much longer. "They all died," Millerd said. "They suffocated. They need a lot of oxygen."

Proponents of fish farming argue that it is the most reliable way to meet consumer demand, while reducing pressure on wild stocks. Don Millerd photo

Eventually, with help from the Vancouver Aquarium, he solved the transportation problems.

"We were learning, learning all the time," said Millerd.

But the fish farms were having problems. The water was too warm and they were struggling with algae blooms and other disasters. "They were growing crappy little fish," he said. "Nobody was making any money."

Don Millerd built his second plant at the site of the long-defunct Englewood mill. Don Millerd photo

And then, at the end of the 1980s, the fish farms started moving north where the water was colder. It was about this time that a friend of Don Millerd's called him on the phone.

"The fish farms are all going to be up here north of the Narrows," he said. "And there's a fellow here at Brown's Bay who thinks he has the perfect place for a processing plant."

"I'll be right there," said Don Millerd. He was just four years out of receivership—just starting to get going. It seemed madness to take on more risk. But Millerd had spent a decade on packers that were waiting for the tide at Seymour Narrows, and he knew what a bottleneck it was for southbound commercial traffic. He looked around and decided that, although the fish farms in the area might prefer to control the processing as well as the raising of their fish, they simply didn't have the capital to do both. And he was convinced that this new industry would eventually change everything. In 1988 he started to build Brown's Bay Packing.

His business friends were frankly skeptical. "Jesus, there goes Millerd again," they said. "He's going to go broke again."

But they were wrong. Don Millerd eventually built a second facility at the old mill site in Englewood—the mill that closed and threw Del Buck out of work—and he has never looked back.

"I was watching Norway," he said. "The European smokers were starting to buy their farm fish instead of Pacific salmon. Why? Because every fish was perfect; it was a reliable product. And just as important, they could buy it, smoke it and sell it. They didn't have to hold frozen stock for months. So the cash flow is completely different.

"Nineteen eighty-eight was a peak year for the wild fishery. Great markets. The Japanese economy was booming. So nobody paid much attention to fish farms. But I kept watching them, watching them improve their product and overcome their problems, and I thought—they're going to make it." And he would make it with them.

"The fact of the matter is, the world likes fresh salmon. Fresh not frozen. There's now over a million tonnes of farm salmon produced—and it's preferred. This whole fish-farming thing is part of a coastal transition. It's the transition from hunting and gathering to cultivating the seas, and it won't be controversial 20 or 30 years from now. This is the way it will be 20 or 30 years from now."

"We've got dwindling wild fish and an increasing world population," said David Rahn, "so fish farming is not going to go away. But they need sales in the first world, because farm salmon isn't a Third World product." And they will get those sales. China, an awakening giant, has just joined the World Trade Organization, which opens up an immense new market.

Even now farmed salmon is BC's largest agricultural export product. The US takes 80 per cent of our production, and in the year 2000 it had a $320-million wholesale value. "But it's not an industry that's making money," David Rahn said. "At today's prices these companies are losing it. The wholesale price has been plummeting due to a world oversupply, but these guys will sell for whatever it takes to get a global share of the market—to get a presence in the international market. And they have deep pockets."

"The Norwegians control the fish farming here," Don Pepper said, "and they have economies of scale. So they lose $10 million here. So what? They make it up somewhere else like Chile where the cost of labour is low. The goal is to create a monopoly.

"There's a lot of jostling going on between the big boys right now. Chile has flooded the market in Japan and screwed the Americans because the Japanese aren't buying Alaska sockeye. So the Americans are yelling 'dumping!' Now Chile has got that sewed up they're trying to get into the European market, but they're being blocked because Norway controls that. The bit players are maybe Iceland and the Faeroe Islands and BC."

Even for the bit players there will be big money, BC fish farmers believe. The question is, what will be the real costs of this prosperity? "There are direct benefits," said Don Pepper, "and we know the direct costs. But the indirect costs are yet to be measured."

A slight young PhD candidate at UBC wants to be one of those who measure the indirect costs. In her short life she has lived under two widely differing regimes—communism and capitalism—and that has given her greater than average perspective (and, incidentally, a reluctance to be identified). She's hoping for enough funding to study the impact of fish farms on the BC economy. She has already spent two years in Norway and recognizes that there are lots of environmental and ecological problems. "But I want to look at the economic costs," she stressed. "I want to find out if fish farms are profitable for society as a whole. Or are they just profitable for private individuals? We don't know that yet."

With a subject like that for her thesis, she's unlikely to receive government funding. For despite the examples of other countries, the environmental degradation that has already taken place and the threats to our native species, Canadian politicians have given fish farms their blessing. A November 12, 2002, story in the *Vancouver Sun* quotes Inka Milewski, a marine science advisor to the Conservation Council of New Brunswick, on the slow government response to mounting evidence of environmental damage. New Brunswick politicians are reluctant to touch a burgeoning industry that employs close to 2,000 people and adds over $230 million a year to the provincial economy.

In the spring of 2002 the BC government announced that it would lift the moratorium on fish farms. But Rafe Mair, the *Vancouver Sun*, the governor of Alaska, First Nations groups, environmentalists and commercial fishermen caused quite a stir, and the government reversed its decision. Odd Grydeland, president of the BC Salmon Farmers Association, was not discouraged by the postponement. "We've been waiting for seven years for the lifting," he said. "We can wait a few more days."

As it happened, he had to wait almost five months. Then, having given everyone the opportunity to vent their anger, the provincial government did what it had always intended to do. On September 13, a *Vancouver Sun* headline read, "BC to lift moratorium on salmon farming: Documents indicate industry will police itself." It was a cynical end run around the public's voiced concern.

That same day Environment Canada issued an ocean disposal permit to Grieg Seafood, a fish farm in Esperanza Inlet. The permit allowed the farm to dump up to 2,800 tonnes of fish about 30 kilometres off the West Coast. Dixie Sullivan, a program scientist with the ocean disposal section, was quoted as saying that this was "an emergency situation." She explained that dead fish put more weight on the fish-farm nets than swimming fish do. She said that something had to be done to prevent live fish from escaping or, worse still, the whole mess fouling the inlet.

Meanwhile a 2002 website bulletin from State of the Environment Norway confirmed that Norwegian fish farmers continue to nuke their watercourses with rotenone in order to control sea lice. But now even this isn't enough to eradicate the parasite. Of the 16 watercourses treated, sea lice have reappeared in five of them.

Unlike the Norwegians, Canadians such as Bill Cranmer and Alexandra Morton and Billy Proctor and John Volpe are battling not only diseases and parasites but also the incursion of an exotic. They fear that aquaculture has the potential to destroy the wild salmon themselves—and with them a way of life.

In November 2002 a petition was being circulated among scientists calling for the resignation of Donald Noakes, head of aquaculture and director of the Pacific Biological Station. It stated that Noakes had failed to initiate appropriate research into the environmental effects of open-netcage salmon farming, that he had attempted to suppress such research, that his investigation into the 2001 sea-lice outbreak was a deliberate sham and that Dr. Noakes and his staff had repeatedly made misleading statements to the public concerning netcage salmon farming.

That same month the Hon. John Fraser, chair of the Pacific Fisheries Resource Conservation Council, warned both levels of government that wild pink salmon in the Broughton Archipelago were in crisis and recommended measures to save them be taken immediately. Members of the Heiltsuk Nation at Bella Bella said that they would go to jail, if necessary, in their fight to keep fish farms out of their area. By January 2003 John van Dongen, BC's Minister of Agriculture, Food and Fisheries, announced that he would be stepping aside because of a police investigation into his handling of an aquaculture file (he was reinstated in May when the Crown said it would not file criminal charges). And then in February the CBC aired a damning indictment of BC's fisheries management on the program *Disclosure*. It revealed that the DFO was aware that its study of sea lice was the cover-up Otto Langer said it was. It outlined, as well, the case of Stolt-Nielsen, the biggest fish-farm company on the Pacific coast. The company was being examined by environment officials for its second major fish escape in a year, but executives—tipped to the investigation by van Dongen—fired off a stern letter to the appropriate BC ministers and the investigation was dropped.

In the corridors of power, Don Pepper's "Duke of Plaza-Toro" is increasingly harassed. He and his friends escape this media onslaught by going out to lunch more frequently than ever, for there are representatives of offshore oil drillers hosting them now, as well as multinational fish farmers and timber companies.

And, if Don Pepper is right, they remain unconcerned. "You can have all the fights and battles and scandals you want," he said, "but still—somewhere, somehow—deals are quietly made."

Canadian Alliance MP John Cummins agreed. "The aquaculture industry has both levels of government in its hip pocket. It's powerful, it's wealthy and it does not tolerate criticism. Even a call for regulations is an attack on it."

Now that we are in danger of losing our wild fish, we have come to appreciate them. Those superb Nimpkish salmon that the Japanese consumer bought at the turn of the last century are being recognized for what they are: a luxury food. Restaurants reviewed in *Gourmet* magazine stress that they serve only wild salmon, perhaps because the website for the Chef's Collaborative urges its members to do so. The chefs are boycotting the aquaculture industry, an industry they say spreads diseases to wild stocks. Their website reports that a good-sized salmon farm produces the equivalent amount of sewage as a town of 10,000. As responsible citizens of the world, they recommend line-caught Alaska salmon or, as a second choice, closed-pen, farm-raised salmon.

On the BC coast Don Pepper fishes for pilchards and sells his prime product to C, an upscale Vancouver restaurant that pays top price for them and thinks they're well worth it. And then there are fishermen like the Hawkshaws, who work out of Prince Rupert. They have developed a net called a "tanglenet" that results in unblemished salmon; once aboard, the fish are kept live in tanks that recirculate sea water. They are only butchered when ready to be shipped. They, too, sell to this same select restaurant. With all this care and attention and extra cost people are trying to prove that salmon are not just another commodity.

C H A P T E R 1 3

High Boats:
Moresby III — A.W. Joliffe (BCP)
Twin Sisters — James Sewid (ABC)

Pioneer Journal, September 4, 1957

In 1995 Ken Lands saw a notice in the *Westcoast Fisherman*. It was an open invitation from the provincial government to families and companies that had been in the fishing industry for 100 years to come forward and receive a centennial fisheries award. Ken Lands is a member of the Huson family on his mother's side; his mother is David Huson's cousin. By 1995 the Huson family had been involved in the fishing industry for almost 120 years, and in due time, Ken Lands, his uncle, his mother and his wife attended a ceremony at the Britannia Heritage Shipyard in Steveston, where they were presented with a bronze plaque and congratulated by David Zirnhelt, then the Minister of Agriculture, Fisheries and Food. They were one of only seven families honoured that day.

Now David Huson, a member of that family, surveyed the early morning waters of Port Harvey from the deck of the *May S.* and decided that the sunny days were over. It was as if someone had entered a pleasant room and turned out the lights. The peacock colours of the last week had become shades of gunmetal; there was cloud low over the water and rain spattering on the cabin roof.

"Guess we might as well get goin', eh." He reached in the wheelhouse window for the tide book. "It'll take us five or six hours to the Narrows. And we're goin' with it so if we're a bit late it doesn't matter."

Gusts of wind tumbled around them as they hauled the anchor up and headed into Johnstone Strait. In September 1950 the fisheries inspector, flying over the area in a patrol plane, counted 341 gillnet boats fishing in Johnstone Strait between Port Neville and Growler Cove. A year later the little local paper reported "a phenomenal number" of boats in the straits. An average of 82 seine boats and 800 gillnetters dotted the water; the paper said that at night "their myriad of gleaming lights bobbing around looked like the lights of a big city." The masters of the Union Steamships, already required to perform feats of miraculous seamanship on a daily basis, now had to add the challenge of threading their way through this barrier of boats and nets.

But on the day that David and Barrie journeyed down Johnstone Strait there was no such congestion. They were alone on that slate-grey stretch of water, alone until the dolphins appeared. Suddenly they burst out of the sea and surrounded the *May S.* Like torpedoes they shot toward her leaving trails of foam until they were racing along three and four abreast on both sides, tight against the hull. Singly and in perfectly choreographed groups of four or five they leapt out of the water. For hours they followed the boat—little grey dolphins that were new to this part of the coast. Some joined the game, some left, but all the way to the Narrows they were there, shooting along on either side of the *May S.*

"California dolphins," David said. "Never seen them this far north until we got that El Niño. Get your camera and put it on flash; that's all you have to do to get their attention. They know you're takin' a picture, it seems. They'll come alongside and sort of look up at you. They're seein' us off."

Roughly halfway down its length the east coast of Vancouver Island bulges outward. The apex of this triangular protuberance is Chatham Point, and it's here that the southbound mariner makes his decision. Will he travel south via the Yucultas, Okisollo and Hole in the Wall, or through Seymour Narrows? In each of these passages the tides that pour around Vancouver Island from the north and from the south meet in narrow channels, creating tide races that run at up to 18 knots and are only negotiated after close attention to the tide book. There are advantages and disadvantages to each route. In bad weather the twisted passages that form the first option offer shelter. But if the weather is fair the straightforward route to Seymour Narrows is favoured by exhausted skippers who can put a man on the wheel with instructions to follow the shore of Vancouver Island—and then lie down and get some rest. "Otherwise in the Yucultas or Okisollo they don't know where they're at," said David.

"We're goin' to be an hour late for the ebb in the Narrows," David said. "So what do you think? We'll get a bit of a ride but I think it'll be okay. Maybe get some bread and milk at Campbell River."

"Whatever."

"Wonder how much fuel we've got, too. I think I put 500 gallons in her but I can't remember where I fuelled up. You know how you get when you get old, eh. If I fuelled her up in town we might be in trouble but I have a feeling I topped up at the Bay in June."

"Well, we've been travelling a fair bit this week so I hope there's enough to get through the Narrows. Stupid old man!"

"That's right, eh. Does your wife call you that, too? Good thing they didn't know us when we were young bucks, eh?"

It was starting to rain. The surface of the water was sliding and the *May S.* sliding with it.

"Here," said Barrie. "You better take the wheel."

David held the wheel with one hand and picked up the binoculars with the other. He stared ahead briefly and then plopped them down. As the *May S.* lay on her side he spun the wheel hard to port, and then hard to starboard, and then hard to port again as the sea boiled up beneath them. Water was coming at them from several directions now. The *May S.* rushed toward the Vancouver Island shore and then, ignoring the fact that her wheel was hard to starboard, she swept across toward Quadra Island. "Wouldn't want to come through here too much later," David said, hauling the wheel over hard again. In the tide boils the *May S.* heeled and slid broadside. "Used to be pretty haywire in here before they took out Ripple Rock."

"Remember the time you and I went up from Vancouver with your granny?" Barrie asked. "We were tied up waiting for high-water slack in the Yucultas and we pushed that 30-foot log off the beach?"

"And it got swallowed up in a whirlpool." David was laughing. "I wish you coulda seen your face."

They swept past Duncan Bay.

"It's pretty shallow in front of Campbell River," David said. "But they treat us pretty good; don't charge us for overnight."

Ahead lay the drenched town of Campbell River. "Look at that weather, eh. I swear every time I've tied up here it's raining. Every damn time."

In the rain they walked up the dock. By the time they got back with their plastic bags, the wind, even there behind the breakwater, was pulling at the *May S.* She tugged at her lines like an obstinate pony on a tether.

"I think we'll just put an extra spring on there later," said David, as they clambered over the bulwarks.

In the galley they had spread their purchases on the table and were starting to put their meal together when the sound of an engine reversing overwhelmed the music from the radio. David snapped on the deck lights and stepped outside. In the gathering darkness a little halibut boat nudged alongside the *May S.* The deckhand who heaved the head line to David was a young woman who looked, he calculated, about two years pregnant. Her dripping hair was plastered to her head.

"You're lookin' kinda' wet," he said. "Better come on over and eat with us."

There were only the two of them on board—a couple who had been fishing up north all summer. Barrie poured coffee and they settled around the galley table and swapped fish stories. The oil stove threw out waves of heat. Out in the dark, rain slashed away at the cabin roof.

Despite the coffee Barrie slept well. He woke to the measured voice of Environment Canada's weather report. "The small-craft warning has been upgraded to gale warning for all inland waters including Georgia Strait. Sisters Island—winds southeast 35, gusting to 45, visibility …" David was leaning against a window frame smoking a cigarette.

"Seas four to six feet," he said.

"That means six to eight."

"Ebb tide from the Narrows into a southeaster."

"Where's that boat that was alongside?"

"She left about an hour ago. The guy said he wanted to get across and behind Texada before it picked up too bad. I'd kinda like to get goin', too." David went into the cabin and got his yellow rain jacket off the wall. "I don't feel like sittin' here for two days. Know what I mean?"

"Maybe stick our nose out and see what it's like," said Barrie.

Halfway out the wheelhouse door, one foot over the threshold, David poked his head back inside. "When we get outside there, watch yourself on that stove, eh."

Heading to Victoria, the Campbell River and Cape Mudge people say that it's shorter to go from Cape Mudge over by Mitlenatch Island to Powell River and down behind Texada Island. The Alert Bay people say it's shorter to go straight down the outside. "It depends on who you're with," David said, and headed straight down the outside.

Out in the tide rip opposite Cape Mudge the *May S.* hurled herself into the seas. She had been doing things like this for 73 years—and she had been lucky. She'd always had a captain who knew what he was doing, as she did now. Confronting each wave, she reared up and then dropped into the trough with a thud, explosions of white water bursting over her bows and sending scud flying over her wheelhouse.

In the galley, Barrie, fighting for balance, staggered from the table to the stove, breaking eggs into the frying pan and buttering toast, timing every movement with that of the boat. Finally he piled fried eggs, bacon, last night's leftover potatoes and a pile of toast onto a plate and pushed out the galley door. Waves hissed and reared around the heaving deck and the exhaust whipped astern as if shot from a gun. Barrie was halfway up the ladder when the *May* slammed down into a trough, almost loosening his grip on the metal rail. Spray arched over the wheelhouse, washing a portion of David's breakfast away with it. David surveyed the remains.

"So now you'll go down in a nice warm galley and have yours," he said.

"I don't know what you figured," said Barrie, "but it's a damn sight nastier out here than I thought it would be."

"Aw, it's not all that bad," said David, but the radio seemed out to prove him wrong. As if on cue the maydays started. Somebody radioed that they were in trouble up on a rock. On this particular morning hitting a rock didn't seem to constitute the highest priority. The Coast Guard, busy fielding calls, put this one in sequence. "Can't get to you right now," the operator said. "You'll have to wait."

And then the men hanging on behind the *May S.*'s dodger heard a message that rivetted their attention. It was their friends on the little halibut boat. The man and his very pregnant wife were out in that turmoil somewhere—and they were foundering.

"Here, take the wheel for a minute while I take a look." David switched the range rings of the radar down to eight kilometres and peered at it intently. "I can see a bunch of shit on there but it's hard to tell what I'm lookin' at. Here, have a look and see if you can figure it out."

He scrabbled around behind him looking for the field glasses and slid one of the Lexan windows open trying to get a look ahead. All he got was a faceful of rain and spray. He slammed it shut.

"My god," said Barrie, instinctively searching the horizon. "They're right up in front of us somewhere."

"From the way I reckon, they must be somewhere off the north end of Mitlenatch."

There was a tug calling in to offer assistance now, and in moments, the Coast Guard's giant twin-rotor Labrador was hammering away overhead.

Barrie is not phlegmatic. "Nobody knows that woman's pregnant," he said, getting more agitated by the moment.

David was staring through the binoculars again. "I think I seen a flare," he said.

"Let me have a look." Barrie grabbed for the glasses. "Dammit, dammit! They would have been fine if they hadn't headed across. Why in hell didn't they run all the way down and go across through Porlier Pass. Dammit. That *is* a flare. I don't know where in hell you were looking but it's about 15 degrees off the bow. It's a location flare for sure but I can't see any lights from the helicopter—no strobes, nothin'." He was yelling now, thinking of the young couple and all the plans they'd shared the night before.

"Don't yell at me," said David. Fishermen fish and fishermen drown; he had 1,000 years of calm in his genes to help him accept that fact.

They both subsided into disgruntled silence. The radio, however, kept up its own rattle of urgent conversation, although in the confusion of continuing calls from distressed mariners it was impossible to follow any one event. The *May S.* continued to rear up and crash down into the trough; each time she climbed a wave the forward mast stay slackened a bit then went drum tight as she dropped into the trough. Each time a small rivulet of water ran down the stay. Suddenly Barrie was rigid with attention. "Quiet for a minute. Listen. They got them. They *got* them. Thank Christ."

It was another hour before they heard the message that made them grin at each other. "The woman you have just rescued would like to know the names of the two men who saved her," said a voice from the radio. "Apparently she's expecting twins and she wants to name her babies after them."

Neither of them spoke as the same voice issued a Coast Guard notice of a hazard to mariners created by the submerged hull of the halibut boat. Only hours before, this hazard had been the young couple's home and business and the key to their future life.

The *May S.* pounded on down the strait. "I'm just thinkin' that if we ever get down around Dodd Narrows and it flattens out a bit we could get somethin' happenin' in the galley," said David. "It's not lookin' too bad out here now. Startin' to clear up a bit, too, from lookin' further down."

From the monotony of this long open stretch of grey combers the voyage went to the other extreme. Dodd Narrows is a deep passage only three or four times the width of a seine boat.

Barrie fumbled for the binoculars and surveyed the entrance. "Remember how we used to con the young guys into steering?" he said. "You'd give them a heading and tell them you were just goin' to get a coffee, and then go and sit in the galley for a couple of hours.

You'd keep an eye out all right but just leave them up there in the rain. They'd be too afraid to leave the wheel to come looking for you. After awhile they caught on."

"We came through here one time with the *May S.* and the *Kitgora* tied together," said David. "We had a new guy on the wheel, his first year on the boat. You've got to have a guy on each boat to go through here but he didn't know that. So we got near the Narrows and we all just snuck away. We all just disappeared and this poor guy thought he was the only one steerin', eh. It was getting narrower and narrower and he's lookin' and yellin' and nobody's comin'. We're all in the wheelhouse, laughin' away."

The *May S.* emerged from the Narrows and worked her way down Stuart Channel.

"We're not in any hurry," David said. "We'll just find ourselves a log boom and tie up here in Chemainus for the night. You okay with that?"

The wind had died and outside the open galley door there was a watery sunset. Ducks paddled past the boomstick where the *May S.* was moored and investigated Barrie's discarded potato peelings. In the galley all was domesticity: bubbling pots and the ling in the frying pan.

"Seems like an awfully long time ago I was on the *Kitgora* with Steve," said Barrie, placing knives and forks on the table's worn Arborite.

"It *was* a long time ago. Steve was 88 when he died and you're no spring chicken yourself."

"Steve was true blue," said Barrie, "just all the things I always thought a man should be."

David dug a cigarette out of a package that had been seriously crushed. He chuckled. "Steve got a hold of me a while ago and he said, 'David, we've got to do this one more thing now before I go.'

"I thought, oh God, he's going to die on us.

"'We've got to ...'

"What's that?

"'We've got to get out in the boat. You know there's nothing like lyin' down in the bunk and listenin' to the engine runnin' and all this stuff.'

Steve Warren aboard the Kitgora. *Barrie McClung described him as 'all the things I thought a man should be.'* David Huson Photo

"God," David said. "He sounds like he's going to go any minute. So I took my little tape recorder down to the boat and I started the engine and left it runnin'—turned the tape on and the engine's just chuggin' away. We put the tape in his Christmas present; we bought him a tape deck. 'Boy,' said Steve, 'that works good.'"

Barrie broke off some pieces of bread and threw them at the ducks. "Steve was a really decent man," he said.

The tentative sunset of the night before became a confident sunrise. David cleaned streaks of salt off the windows with a squeegee. "Nice warm day," he said. "Flat calm. Great day for cruisin'."

69302
302

They slipped through Sansum Narrows, sucking in the perfect day like a restorative. At lunch they stopped the engine and drifted off a mansion of a house on Landsend Road.

"That's a big fancy place, eh," said David.

Barrie was setting out a plate of sandwiches and mugs of coffee on the hatch cover.

"Nothing I would trade for the *May S.*" He picked up a sandwich, leaned back in his chair and regarded the glassy sea. Little scraps of kelp as glossy and golden as amber drifted by. "Wonder what the poor people are doing?" he said.

David's eyes crinkled with amusement. "I was a big shot one time," he said.

"You?"

"Yeah. When my dad quit fishing, John Barry and Bill Bell from his crew bought a limousine. They worked for Bluebird Taxi, eh. One time they came an' picked us all up, my sister and her husband and my dad and me, and we went to the Imperial Inn. The guy with the little hat gets out and opens your door. We all got out. Big shots, eh. John went in and said, 'I've got this party ...' Boy, they got us a table really fast. We did really well but then we had a few drinks and there was stuff falling over and that." He was chuckling. "Then they knew they didn't have the real classy guys there."

Barrie went into the galley for more coffee. "It was nice for my dad, though," David said, serious now. "My mother had MS and it sort of killed my dad in the end—all the work and worry of lookin' after her. But these guys would come along in the limousine. They were happy-go-lucky guys. And they'd take him out for a ride around. They had this little bar in a briefcase so the two in the back could have a good time. They did things like that. They had the taxi but they didn't worry too much about the money."

Off the Oak Bay golf course there was a fresh breeze and a

Opposite: A sailboat, its colourful spinnaker billowing in the wind, passes the May S. *off the Royal Victoria Yacht Club in Oak Bay.* From the collection of Barrie McClung

As motorists wait on either side, the Johnson Street Bridge swings up to admit the May S. From the collection of Barrie McClung

scattering of sailboats dotting the water. Yacht-club boats were racing; turquoise and purple, red and yellow, their spinnakers billowed in the wind. The *May S.*, like the workhorse she was, plodded past these exotic butterflies and followed the curve of the shore past Ogden Point and into Victoria's harbour. Coastal seamen, used to the grandeur of Vancouver harbour, refer to Victoria's as a duck pond. But ringed by lovely old buildings and tasteful new ones, and surrounded by lawns and trees and shrubbery, it substitutes its own gentle charm for Vancouver's industrial strength.

David picked up the radio phone and called the Johnson Street Bridge. "This is the *May S.*," he said. "Can we have a lift?"

They watched as the road traffic was halted. Slowly the whole span of the bridge rose from its eastern end and angled into the late summer sky. Just for us, Barrie said to himself, just for us because we have come all the way from Alert Bay.

"All the times I've gone in and out of here I've only had to pay once," David said. "They make you pay $75 if it's midnight or somethin'."

They moored at the old Coast Guard station just inside the bridge, and then they threw their bags and a couple of salmon, now rock-hard, up on the dock.

The voyage of the *May S.* had come to an end.

It was Barrie's wife, Linda, who answered the phone. She handed it to Barrie. "It's David," she said.

David got right to the point. "Well, I sold my boat," he said. "I sold the *May S.*"

"You what?" Barrie was dumbfounded. "I thought we were bringing her to town …"

"To sell her. Man runnin' a kayak business bought her. With that big back deck it'll work out just right for him." There was a long pause. "It's kinda' hard. The *May S.* was my father's boat. He was so proud of that boat."

Barrie found his voice. "But David, couldn't you …"

"I'm tellin' you, Barrie, I couldn't do *nothing*. I even tried leasin' her last year but them things never work out. I tried to make it work but it wouldn't work. I can't hang on to a boat like that just so you and me can go cruisin' around."

So that's what we were doing, Barrie thought. Cruisin' around. Saying goodbye to Izumi Rock, the Bluff, the Merry-Go-Round. Saying goodbye to the dolphins.

"I seen it," said David. "I told my friends, 'It's comin'. Look at the east coast. Within five years we'll be there.'"

"Oh, shit."

"I thought I was the first fisherman in the world, Barrie. I just loved it. I got in just at the tail end. I fished everything there was to fish. I was just a kid ridin' around."

There was a long silence.

"Oh, shit," said Barrie, softly.